Morphology of the Rat Brain Ventricles, Ependyma, and Periventricular Structures

Bibliotheca Anatomica

No. 21

Series Editor: *W. Lierse*, Hamburg

S. Karger · Basel · München · Paris · London · New York · Sydney

Morphology of the Rat Brain Ventricles, Ependyma, and Periventricular Structures

A. Mitro and *M. Palkovits*

75 figures, partially in color, and 4 tables, 1981

S. Karger · Basel · München · Paris · London · New York · Sydney

BIBLIOTHECA ANATOMICA

No. 17: Microcirculation in Inflammation
Hauck, G., Münster, and *Irwin, J.W.*, Boston, Mass. (eds.)
VIII+148 p., 52 fig., 4 tab., 1980. ISBN 3-8055-3016-1
No. 18: Current Advances in Basic and Clinical Microcirculatory Research
Lewis, D.H., Linköping (ed.)
XII+400 p., 140 fig., 46 tab., 1980. ISBN 3-8055-3042-0
No. 19: Studies of Normal and Abnormal Development
of the Nervous System
Lierse, W., Hamburg, and *Beck, F.*, Leicester (eds.)
VIII+328 p., 178 fig., 11 tab. 1981. ISBN 3-8055-1039-X
No. 20: Recent Advances in Microcirculatory Research
Gaehtgens, P., Köln (ed.)
XIV+730 p., 283 fig., 69 tab., 1981. ISBN 3-8055-2272-X

National Library of Medicine, Cataloging in Publication
Mitro, A.
Morphology of the rat brain ventricles, ependyma, and periventricular structures
(by A. Mitro and M. Palkovits. — Basel; New York: Karger, 1981.
(Bibliotheca anatomica, no. 21)
1. Cerebral ventricles — anatomy & histology 2. Ependyma — anatomy &
histology I. Palkovits, M. II. Title III. Series
ISBN 3-8055-2546-X
W1 B1394 no. 21 [WL 307 M684m]

A joint publication of S. Karger AG, Basel with Akadémiai Kiadó, Budapest

Printed in Hungary.
S. Karger AG, P.O. Box 4009 Basel (Switzerland) has the sole distribution rights for all countries, with the exception of the socialist countries.
ISBN 3-8055-2546-X

CONTENTS

PREFACE VII

LIST OF ABBREVIATIONS IX

INTRODUCTION 1

MATERIAL AND METHOD 5

TOPOGRAPHY OF THE VENTRICLES 7

Localization of the Ventricles in the Coronal Serial Sections of the Brain 8

Localization of the Ventricles in Sagittal Serial Sections of the Brain 30

Localization of the Ventricles in Horizontal Serial Sections of the Brain 30

EPENDYMAL AREAS OF THE BRAIN VENTRICLES 35

Current State of Knowledge about the Ependyma 35

Lateral Ventricle 39

The Third Ventricle 48

Cerebral Aqueduct and Mesencephalic Ventricle 84

The Fourth Ventricle and the Beginning of the Central Canal 90

GENERAL CONSIDERATIONS 99

REFERENCES 101

SUBJECT INDEX 108

PREFACE

In the investigations of the ventricular system of the rat brain in routine practice we relied upon 1. works studying the shape of the ventricular system on casts; 2. atlases of the rat brain as a guide of orientation in periventricular nerve tissue; 3. works dealing with the structure of the ependyma when examining the ependymal lining.

The aim of this work was the description of the topography of the brain ventricles, the ependyma and periventricular neuropil. We suppose that our descriptions will be useful when examining the brain ventricles after various experimental conditions. For such investigations it is essential to compare identical sections of the ventricular wall from various experimental animals designated by a uniform terminology.

We should like to express our gratitude to Professor T. H. Schiebler, Director of the Department of Anatomy, University of Würzburg, for his kind approval of publishing the results obtained while working in his Department. From the Institute of Experimental Endocrinology of the Slovak Academy of Sciences, Bratislava, we should like to acknowledge the help of Alexander Kiss, D.Sc. in evaluating and documenting some parts of the results.

We feel greatly indebted to Professor Szentágothai for providing the opportunity to have this work published by the Publishing House of the Hungarian Academy of Sciences.

the *Authors* A. Mitro
Institute of Neurobiology
Slovak Academy of Sciences
Košice, Czechoslovakia

M. Palkovits
1st Department of
Anatomy, Semmelweis University
Medical School
Budapest, Hungary

LIST OF ABBREVIATIONS USED IN THE TEXT AND FIGURES

a — nucleus accumbens
A — amygdala
AC — aquaeductus cerebri
aD — nucleus accessorius (Darkschewitch)
ap — area postrema
apt — area praetectalis
ar — area retrochiasmatica
A2 — A2 catecholaminergic cell group

c — nucleus caudatus
C — cingulum
ca — cornu anterius of the lateral ventricle
Ca — central canal
CA — commissura anterior
CAE — capsula externa
CAI — capsula interna
CC — corpus callosum
CFV — commissura fornicis ventralis
cgl — corpus geniculatum laterale
ci — cornu inferius of the lateral ventricle
CI — colliculus inferior
cL — central part of the lateral ventricle
CO — chiasma opticum
cod — nucleus cochlearis dorsalis
cp — nucleus caudatus–putamen
CP — commissura posterior
CS — colliculus superior
cT — connecting part of the third ventricle
C2 — C2 adrenaline-containing cell group

DP — decussatio pyramidis
dT — dorsal part of the third ventricle

f — fastigium
F — fornix
FG — fasciculus gracilis
FH — fimbria hippocampi
FI — foramen interventriculare
FLM — fasciculus longitudinalis medialis
FR — fasciculus retroflexus
FS — fimbria septi

GD — gyrus dentatus
gp — globus pallidus
gr — nucleus gracilis

H — habenula
HI — hippocampus
hl — nucleus habenulae lateralis
hm — nucleus habenulae medialis

i — nucleus interpositus cerebelli

l — nucleus lateralis cerebelli
lc — locus coeruleus
LV — lateral ventricle

m — nucleus medialis cerebelli
MB — mamillary body
me — median eminence
mi — massa intermedia of the thalamus
MO — medulla oblongata

na — nucleus arcuatus
nco — nucleus commissuralis
ndm — nucleus dorsomedialis
nha — nucleus hypothalamicus anterior
nhp — nucleus hypothalamicus posterior
nic — nucleus intercalatus
nist — nucleus interstitialis striae terminalis
nl — nuclei lineares
nmm — nucleus mamillaris medialis
nmx — nucleus mamillaris medialis, pars lateralis
npe — nucleus periventricularis hypothalami
npmd — nucleus praemamillaris dorsalis
npf — nucleus parafascicularis
npv — nucleus paraventricularis
nrp — nucleus reticularis paramedianus
nsc — nucleus suprachiasmatis
ntd — nucleus tegmenti dorsalis
ntdl — nucleus tegmenti dorsolateralis
nts — nucleus tractus solitarii

nvm — nucleus ventromedialis
nX — nucleus originis dorsalis vagi
nXII — nucleus motorius nervi hypoglossi

O — obex
ovlt — organum vasculosum laminae terminalis

P — pons
PCS — pedunculus cerebellaris superior
PG — pineal gland
ph — nucleus praepositus hypoglossi
pom — nucleus praeopticus medialis
pop — nucleus praeopticus periventricularis
pos — nucleus praeopticus suprachiasmatis
pp — posterior portion of the lateral ventricle
ps — pituitary stalk
pt — nucleus parataenialis thalami
ptv — nucleus periventricularis thalami

rd — nucleus raphe dorsalis
re — nucleus reuniens
rgi — nucleus reticularis gigantocellularis
Ri — recessus infundibuli
RI — recessus inframamillaris
RL — recessus lateralis
RO — recessus opticus
RP — recessus pinealis
rpoc — nucleus reticularis pontis caudalis
RT — recessus triangularis

s — nucleus suprageniculatus
S — subiculum
sa — nucleus salivatorius
SCO — subcommissural organ
sd — nucleus septalis dorsalis
SE — septum
sf — nucleus septalis fimbrialis
sfo — subfornical organ
SGC — substantia grisea centralis
SGCd — substantia grisea centralis, pars dorsalis
SGCl — substantia grisea centralis, pars lateralis
SGCv — substantia grisea centralis, pars ventralis
si — nucleus septalis intermedius
sl — nucleus septalis lateralis
sm — nucleus septalis medialis
SM — stria medullaris thalami
st — nucleus septalis triangularis

tad — nucleus anterior dorsalis thalami
tav — nucleus anterior ventralis thalami
td — nucleus tractus diagonalis
TD — tractus diagonalis
TH — thalamus
tl — nucleus lateralis thalami
tlp — nucleus lateralis posterior thalami
tm *m* — nucleus medialis thalami
tml — pars lateralis of the nucleus medialis thalami
TO — tractus opticus
TP — tegmentum pontis
tr — nucleus reticularis thalami
TS — tractus solitarius
TSTH — tractus striohypothalamicus

V — vermis cerebelli
vl — nucleus vestibularis lateralis (Deiters)
vm — nucleus vestibularis medialis
VM — ventriculus mesencephali
VMS — velum medullare superius
vs — nucleus vestibularis superior
vsp — nucleus vestibularis spinalis
vT — ventral part of the third ventricle

IV. — fourth ventricle
VII. — nervus facialis

INTRODUCTION

The walls of the brain ventricles and the central canal of the spinal cord are covered by a variety of cells forming together an epithelium-like lining called ependyma. The origin of the term is somewhat obscure. As early as 1841, the word "ependyma" was used in Valentin's edition of Soemmering's (1841) "Textbook on Human Anatomy" to denote the membrane covering the walls of the cerebral ventricles (cit. by Fleischauer, 1972).

One of the first detailed descriptions of the ependyma was published by Purkinje in 1836, in which he described cilia and their basal bodies in the apical part of the ependymal cells. The discovery of the choroid plexus is also linked with his name.

The epithelial characteristics of the ependyma, i. e. the existence of an epithelium inside the central nervous system was described and proved by Kölliker in 1896. An extensive study on the structure of ependyma was written by Studnička (1900) followed by a number of detailed descriptions of the ventricles and the ependyma in various species and in human (Agduhr, 1932; Pesonen, 1940; Horstmann, 1954; Fleischauer, 1957; Schimrigh, 1966; Böhme and Bodo, 1967; Malinský, 1968; Leonhardt, 1969; Levinger, 1971; Scott et al., 1972; Levinger and Kedem, 1974). In the last decade a number of studies further confirmed that there are significant regional variations in the structure of the ventricular ependyma (Fleischauer, 1957; Blinzinger, 1962; Leonhardt, 1966, 1968; Takeichi, 1966; Colmant, 1967; Srebro, 1967; Rodriguez, 1969; Rodriguez, 1970a, b; Westergaard, 1970; Ogata et al., 1972; Mitro, 1975; Kiss and Mitro, 1976; Hetzel, 1977; Mitro and Kiss, 1977). Further structures related to the ependyma such as the subependymal elements and supraependymal structures have also been described (Leonhardt and Lindner, 1967; Blakemore and Jolly, 1972; Privat and Leblond, 1972; Scott et al., 1973).

In spite of the great number of papers dealing with the ependyma there are many basic terms that are still vaguely defined. Among these the first is the ependyma itself. According to Knowles (1972), "the term ependyma is generally used to denote the lining of the inner walls of the brain and spinal cord bordering the ventricles or spinal canal".

According to the description of Leonhardt (1969), the term ependyma is used in a wider sense including other elements than the ependymal cells such as nerve fibres and nerve cells terminating in or located among the ependymal cells, and taking part in the formation of the ependymal lining. In our interpretation, the ependyma as an integral constituent of the central nervous system contains ependymal cells in one or more layers with other intraependymal structures such as the intraependymal nerve elements, intraependymal capillaries, labyrinths, etc.

The ependymal cells, having various shapes, are arranged in one or more layers. Flat, cuboidal and columnar (cylindrical) ependymal cells can be distinguished. Some of these, possessing basal processes varying in length, are termed tanycytes (Horstmann, 1954). The majority of the ependymal cells, however, have no basal extensions and are separated by a basement membrane of the cell facing the underlying neuropil.

The interface between the ependymal lining and the neuropil is mostly irregular. This is due to tanycyte processes intermingled with the neural elements (dendrites, axons) of the neuropil. Furthermore, periventricular neurons and subependymal glial cells are also intermingled with the ependyma causing the irregularities of the interface.

The cell types and thickness of the ependymal layer and of the sub- and supraependymal elements vary greatly in different regions of the ventricular wall. Unfortunately, our present knowledge is by far insufficient concerning either the fine structure of the ependyma or that of the regional characteristics. This is probably a consequence of a number of problems encountered when evaluating the ependyma:

1. The relative scarcity of systematic morphological studies of the brain ventricle system;
2. Insufficient delineation of the investigated regions;
3. Inadequate topographic description of the regions investigated;
4. The lack of a uniform classification of various ependymal elements. All these cause serious difficulties when trying to compare observations by different authors.

The present paper is focussed on the following studies:

1. Topography of the ventricle system in rat brain with special regard to its three-dimensional coordinates;
2. Measurements to determine the dimensions of various ependymal areas including their percentage of the total ventricle surface;
3. Detailed description of various ependymal areas;
4. Studies on possible relations between different ependymal areas and periventricular structures.

Placing the ependymal regions in a coordinate system, a uniform classification and probably a uniform interpretation of the ependyma can be developed that would be essential for the comparison of results published by different authors.

The fact that neuropeptides and neurohormones are found in measurable amounts in the cerebrospinal fluid (CSF) (Knigge et al., 1973; Joseph et al., 1975) calls attention to the significance of both CSF and ependyma in the cerebral transport of these substances. This transport should be carried out by the active or passive participation of the ependymal cells (Knigge and Joseph, 1974; Scott et al., 1974c).

Ependymal cells of the infundibular recess have received the most attention since Löfgren (1959a, b) formulated his hypothesis that an exchange of secretory substances between the hypothalamus and the anterior pituitary takes place via the infundibular recess.

Knigge and Scott (1970) pointed out that the median eminence might be a region of great structural and functional importance with respect to the transfer of regulatory information from the brain to the pituitary gland. Structurally, these functions of the median eminence are localized in its inner ependymal, middle fibrous, and outer palisadic-contact zones (Röhlich et al., 1965; Rinne, 1966). Matsui (1966) found that most of the ependymal and glial processes have direct contact with the basement membrane of primary plexus capillaries. The occurrence of exocytosis at the perivascular membrane of the vascular processes indicates a release of substances into the blood (Wittkowski, 1968, 1972). In the subependymal and internal zones of the rat median eminence nerve fibres and vascular processes of ependymal and subependymal cells form neuroglial bundles. Numerous neuroglial contacts were found between nerve fibres and vascular processes of glial cells, the vascular processes containing polymorphous granular inclusions. In bilaterally adrenalectomized rats the number and diameter of granules increase (Wittkowski, 1973).

Oota et al. (1974) described both neurosecretory axons and ependymal (hypependymal) glial processes terminating on the capillary walls of the primary plexus of the hypophysial portal vessels in the median eminence of several vertebrates. The ratio of axon terminals to ependymal processes, however, varies considerably in different species. Knowles (1974) described two ways — long- and short-loop feedback — by which the cerebrospinal fluid might provide a medium for endocrine regulation. Both postulations require the presence of hormones in the CSF.

Scott et al. (1974b) investigated the median eminence of the rat by means of light and electron microscopic autoradiographic analysis after the intraventricular infusion of ^{3}H-dopamine. This study demonstrated that ^{3}H-dopamine, once infused into the mammalian cerebral ventricular system, is rapidly and selectively absorbed by tanycytes and transported to the contact zone within 5 min after ventricular infusion.

Kobayashi and Matsui (1967) found monoaminergic synapses in the posterior region of the rat and pigeon median eminence that may participate in the regulation of functional activity of the capillaries of the primary plexus. Silverman et al. (1973) demonstrated the amino acid and thyroxine transport capacity of organ-cultured median eminence. They believe that the ependymal components of the median eminence are responsible for the major portion of amino isobutyric acid and thyroxine accumulation. Kroon and Goossens (1974) studied granules in glial and ependymal cells of the tuberal hypophysiotropic region of the rat hypothalamus. The number of granulated glial cells increases in lactating rats. Uemura and Kobayashi (1977) have shown that the tanycyte transport of substances from the ventricular fluid to the capillaries of the portal vessels is not essential to maintain gonadotrophic activities of the pituitary gland. The ultrastructure of the ependymal cells of the shark median eminence was studied by Saland et al. (1974). The fine structure of ependymal cells of the crested newt was investigated by Fasolo et al. (1972).

The circumventricular organs (these particular regions and special formations of the ventricle walls covered by special cell structures of ependymal origin) may have extraordinary importance in the transport of various substances by the CSF (Weindl, 1973; Sterba, 1977; Felix and Akert, 1977; Weindl et al., 1978). Thus, it is possible that the ependyma in various parts of the ventricle system, especially the tanycyte ependyma, allows communication between the CSF and blood and between the CSF and subependymal neuropil through its cell layers (Ben-Jonathan et al., 1974; Uemura et al., 1975). For this reason, the description of the exact topography and structural organization of the different ependymal areas are among the most actual tasks of neuromorphology.

MATERIAL AND METHOD

Adult male, Wistar albino rats of 200–250 g from the animal house of Dobra Voda (Czechoslovakia) were used for these studies.

The topography and the structure of the ependyma were studied by both light and electron microscopy.

Light microscopy

The brains were fixed by intraaortic perfusion of 4% formalin solution or by immersion in the same solution. After 2–7 days of fixation the total brains were embedded in paraffin, and cut in frontal, sagittal or horizontal planes. The thickness of the sections was usually 8–10 μm. The so-called serial sections contain every fifth section, or even series of sections without skip.

For light microscopic studies the sections were stained with haematoxylin and eosin, or luxol fast blue, or luxol fast blue and cresyl violet.

Electron microscopy

Some brains were fixed by aortic perfusion with 3.5% glutaraldehyde, the others were immersed directly in the same solution (Karnovsky, 1965). Small pieces of the brain tissue (1–2 mm in diameter) were dissected, postfixed in 4% OsO_4, and embedded in Araldite.

For topographic orientation or for a survey of relatively large ependymal areas, semithin sections were used; 1 μm sections were cut by a Reichert ultramicrotome and were stained with 1% toluidine blue solution. Ultrathin sections were stained with uranyl acetate and lead citrate (Reynolds, 1963) and examined under Zeiss 9, Tesla BS 240, and Tesla 500 electron microscopes.

Measurements on the ventricle ependyma

Complete serial sections from every 100 μm were used for the determination of the exact topography and dimensions of various ependymal areas. Sections of the brain of rats, weighing 200 g, were made as described above.

The shrinkage due to the histotechnical procedure was measured and allowed for the living state as described earlier (Palkovits et al., 1971). The length of the walls of brain ventricles was measured in frontal sections from every 300 μm rostro-caudally (see Fig. 2A). The length of the cross sections of certain ependymal areas was measured by an electronic curvimeter and the rear two-dimensional surface of these areas was calculated from the perimeters of the section profiles and the distances of the sections between the two areas measured actually. The surfaces of various ependymal areas are given in mm^2 and also in percentage of the total surface of the brain ventricle system.

TOPOGRAPHY OF THE VENTRICLES

Macroscopy

The ventricle system of the rat — like in other mammals — can be subdivided into the following parts (Fig. 1).

1. *Lateral ventricles.* They form a couple of cylinders placed in the left and right brain hemispheres. In contrast to the human lateral ventricles, the inferior horn is underdeveloped and the posterior horn practically does not exist in the rat. The elongated anterior horn is followed by a relatively large central part of the ventricle continuing in the inferior horn ventralwards at the level of the caudal portion of the amygdala.

2. *Third ventricle.* It is situated in the sagittal plane divided into a dorsal and a ventral portion by the massa intermedia of the thalamus. Its shape reminds us of an irregular hollow cylinder with processes. Four major recesses can be distinguished: supraoptic, infundibular, inframamillary, and pineal (or habenular) recesses.

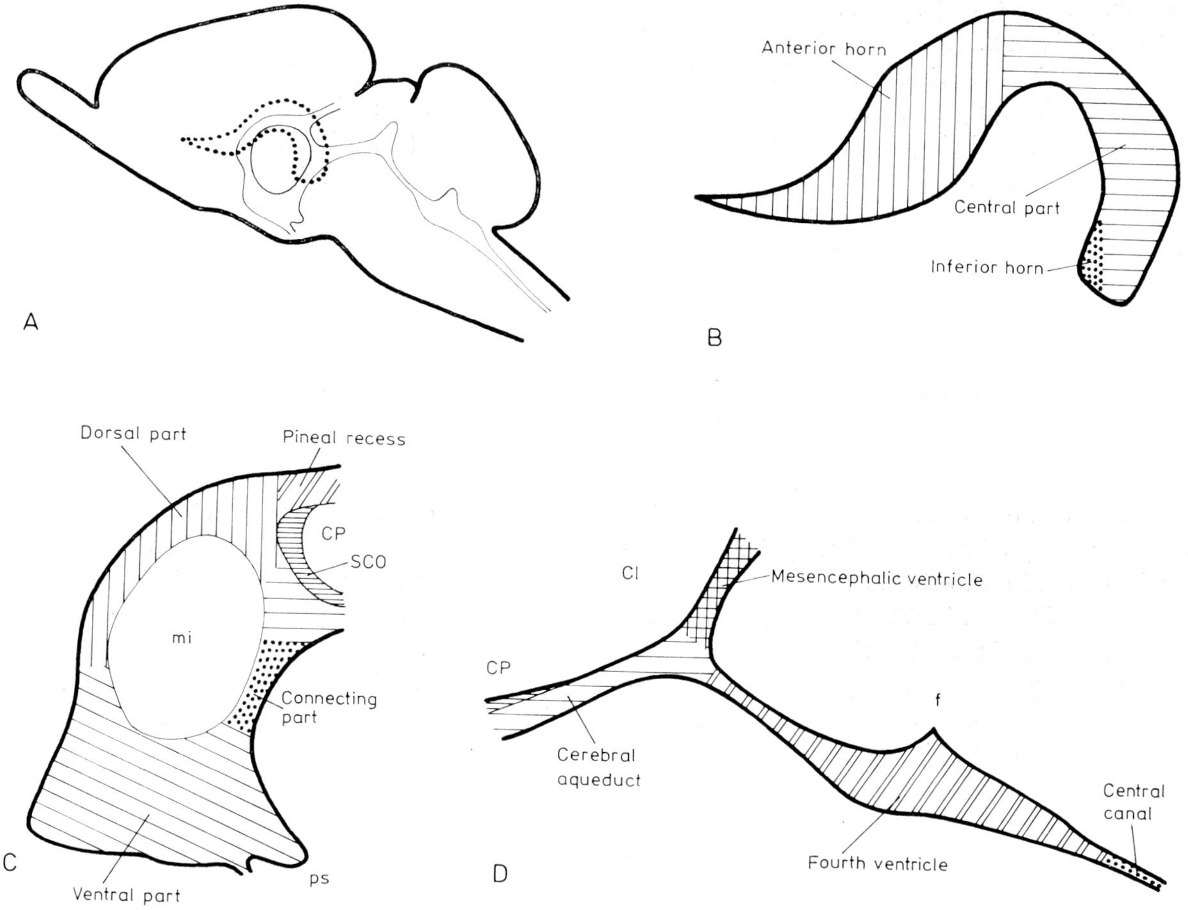

Fig. 1. Schematic drawings of the brain ventricles in sagittal view (**A**) and in detail: **B** — lateral ventricle; **C** — third ventricle; **D** — ventricle system in the lower brain stem

3. *Cerebral aqueduct.* It connects the third and fourth ventricles through the mesencephalon. In the central part of the aqueduct a number of small recesses can be observed on its basal wall, and a large one in its caudal portion.

4. *Mesencephalic ventricle.* It infolds dorsally into the inferior collicle. This is frequently referred to in the literature as the mesocoelic recess or the recess of the superior colliculus.

5. *Fourth ventricle.* The fourth ventricle is located above the pons and medulla oblongata, between the aqueduct and central canal. Its floor (fossa rhomboidea) is composed of the dorsal surface of the tegmentum, while its roof is more complex. The ependyma of the superior medullary velum can be followed up to the fastigium from where, as a constituent layer of the tela choroidea of the fourth ventricle (lamina epithelialis), it forms an irregular fine roof. At the lateral corners of the rhomboid floor the fourth ventricle communicates through the lateral apertures with the external cerebrospinal fluid (CSF) spaces.

6. *Central canal.* At the level of the obex the fourth ventricle continues caudalwards in the central canal. Two parts, a medullary and a spinal one can be distinguished topographically. The area postrema, a specialized region, belonging to the circumventricular organs, is found at the rostral end of the central canal. At the level of the pyramidal decussation the medullary part passes through to the spinal part without any structural changes.

Surface areas of the brain ventricles in mm^2 and in percentage of the total ventricular surface are presented in Table 1. The order of the various parts according to their surface areas is the following: lateral ventricle (39.1%), third ventricle (26.9%), fourth ventricle (22.4%), cerebral aqueduct (8.2%), mesencephalic ventricle (1.8%), medullary part of the central canal, and the circumventricular organs (0.8% each).

Table 1

SIZE OF SURFACE AREAS OF THE BRAIN VENTRICLES

	Surface areas	
	mm^2	%
Lateral ventricles	30.89	39.1
Third ventricle	21.28	26.9
Cerebral aqueduct	6.47	8.2
Mesencephalic ventricle	1.42	1.8
Fourth ventricle	17.73	22.4
Central canal (medullary part)	0.64	0.8
Circumventricular organs	0.67	0.8
	79.10	

LOCALIZATION OF THE VENTRICLES IN THE CORONAL SERIAL SECTIONS OF THE BRAIN

The anterior horn of the lateral ventricle is the rostralmost tip of the ventricle system appearing in coronal sections 3.0 mm rostral to the level of the bregma in 200 g rats* (Fig. 2A, B).

The ventricle is located first laterally to the fibres of the genu of the corpus callosum, more caudally it is shifted to the ventral surface of the corpus callosum. At A0.3 the anterior horn of the lateral ventricle is still the only ventricle on the sections located between the septum and the caudate nucleus. At this level, the supraoptic recess of the third ventricle appears immediately above the base of the brain (Fig. 2/10). This rostral edge of the third ventricle is surrounded by the cells of the organum vasculosum laminae terminalis.

Rostral to the bregma, the lateral ventricle is of a perpendicular position with a slight outward curvature of the ventral tip. Near the bregma level it becomes bent, having a curvature along the lateral edge of the septal fimbria (Fig. 2/11). At about 0.6 mm behind the bregma

* Coordinates in the coronal sections are given in millimetre rostral (A = anterior) or caudal (P = posterior) to the plane of the bregma (0 = zero level = 0 mm).

the lateral ventricles communicate with the third ventricle through the interventricular foramen (Fig. 2/13). This level corresponds to the caudal border of the anterior horn. From here the central part of the lateral ventricles is found between the hippocampal fimbria and the caudate putamen or fibres of the internal capsule.

The supraoptic recess continues into the cavity of the third ventricle at about the bregma level. (Fig. 2/11). Immediately caudal to this region, a small recess (triangular recess) can be observed between the anterior commissure and the column of the fornix (Fig. 2/12). Behind the interventricular foramen the third ventricle is divided by the massa intermedia of the thalamus into a dorsal and a ventral part.

The dorsal part is bordered by epithalamic structures such as the stria medullaris, periventricular thalamic and medial habenular nuclei, and is covered by the tela choroidea of the third ventricle.

The thin, midsagittal ventral part of the third ventricle belongs to the hypothalamus. Its base lateralwards above the median eminence becomes triangular. From this part of the third ventricle the infundibular recess penetrates into the pituitary stalk.

Behind the level of the origin of the pituitary stalk from the hypothalamus (P3.5 mm), seven ventricular transections can be observed in one single frontal section of the brain (Fig. 2/24). The third ventricle is divided into the "connecting part" that connects the third ventricle with the cerebral aqueduct and the inframamillary recess penetrating deeply into the mamillary body. Dorsally, between the habenula and subiculum, the largest cross section of the dorsal part of the third ventricle can be observed. The section profiles of the central part, respectively, the inferior horn of the lateral ventricles, are found on the lateral surface of both the ventral and dorsal hippocampus.

The frontal section of the posterior commissure serves as a landmark for the rostral edge of the mesencephalon and also for both the cerebral aqueduct and the pineal recess of the third ventricle (Fig. 2/26). Both the dorsal and ventral surfaces of the posterior commissure are covered by the specialized ependymal cells of the subcommissural organ which is one of the circumventricular organs. Its cells constitute the floor of the pineal recess and the roof of the aqueduct between P4.5 and P4.8 mm.

Behind the posterior commissure, the aqueduct is shifted ventrally into the centre of the central gray matter (Fig. 2/32). Caudally, the lateral ventricles can still be observed on the lateral surface of the hippocampus as far as P6.3 mm.

The cross sections of the aqueduct are irregular in shape. The "ventral recesses" and the mesencephalic ventricle can be well demonstrated in coronal sections.

The cerebral aqueduct leads into the fourth ventricle about 9 mm behind the bregma (Fig. 2/41). The floor of the ventricle is constituted by the tegmentum of the pons and medulla with a relatively deep sulcus in the midline. The roof, formed by the superior medullary velum, is slightly compressed by the cerebellar vermis as the lateral edge of the ventricle touches the cerebellar white matter. Caudally to the fastigium, which is not so pronounced as in man, the roof of the ventricle is formed by the tela choroidea of the fourth ventricle. At the pontine-medullary border (P11.7 mm) the fourth ventricle communicates with the external liquor space through the lateral aperture (Fig. 2/50). From here the cross sections of the ventricle resemble a triangle in the midline.

In the rat, both the obex and the area postrema are found at the origin of the central canal (P14.0 mm). The relatively narrow lumen of the central canal, bordered by the nucleus of the solitary tract, and the motor nucleus of the hypoglossal nerve, can be followed until the pyramidal decussation.

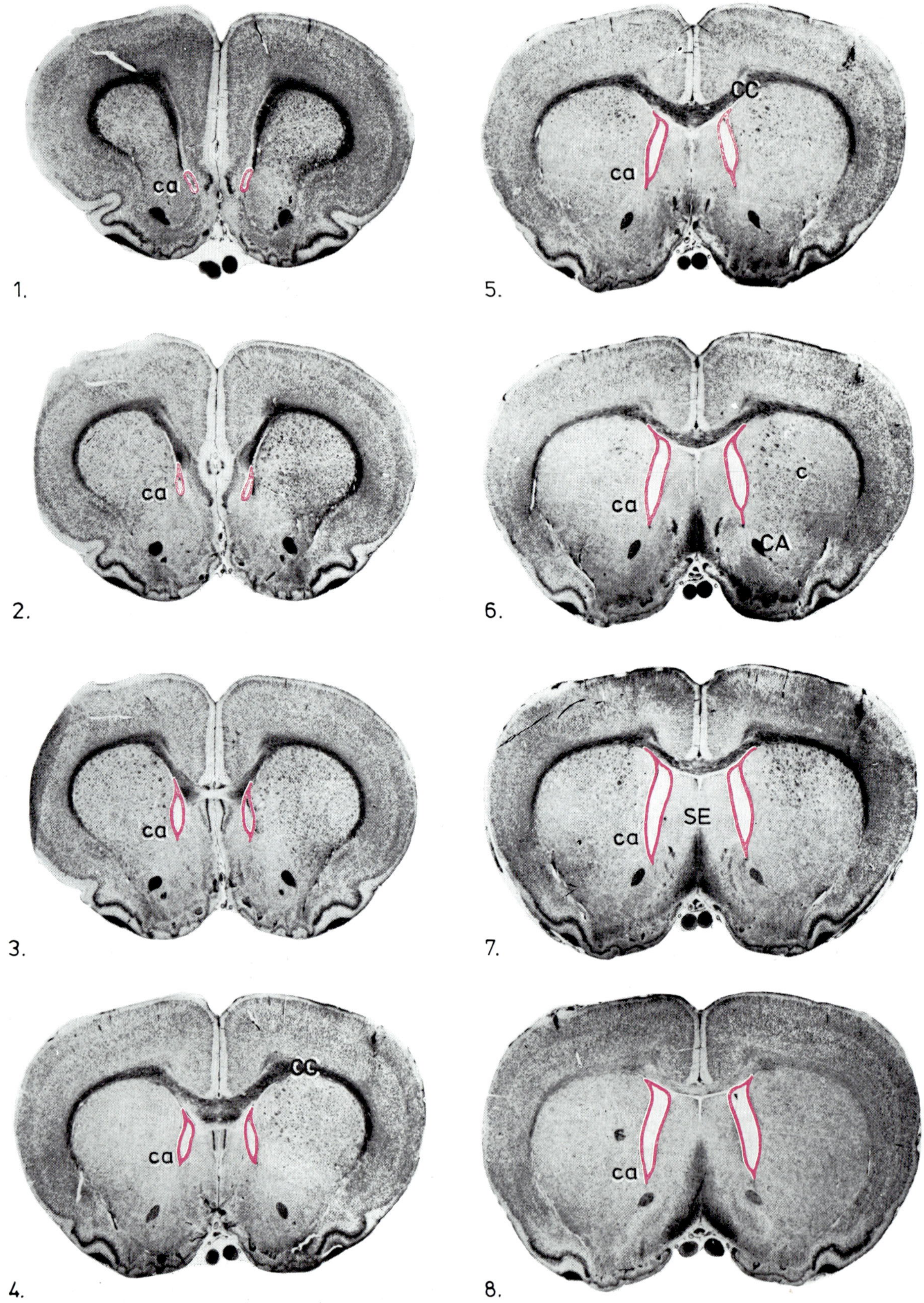

Fig. 2**A**, sections 1–8

Fig. 2. Frontal serial sections of the rat brain. **A** — luxol fast blue cresyl violet staining. 10 μm thick sections at every 300 μm. Magnification: ×10. Rostrocaudal coordinates are given in μm rostral (A) and caudal (P) from the bregma level (0 mm) which corresponds to section 11. **B** — schematic representations of the frontal serial sections from every 300 μm of the rat brain. (Sections correspond to series **A**.) The different parts of the ventricles are indicated by the same colours as in Figs 5, 14, 23, 59, and 68. For abbreviations see list

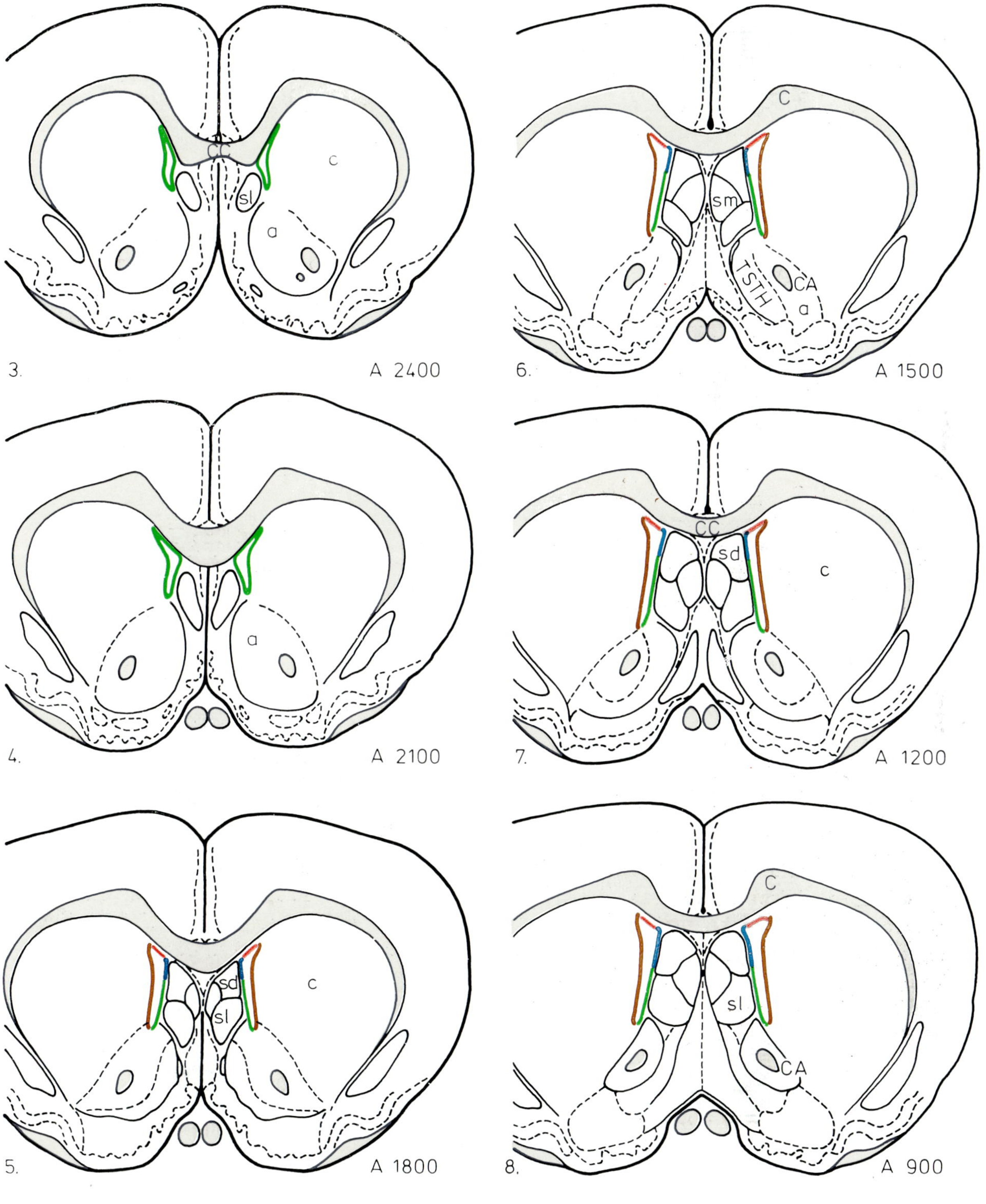

Fig. 2**B,** sections 3–8

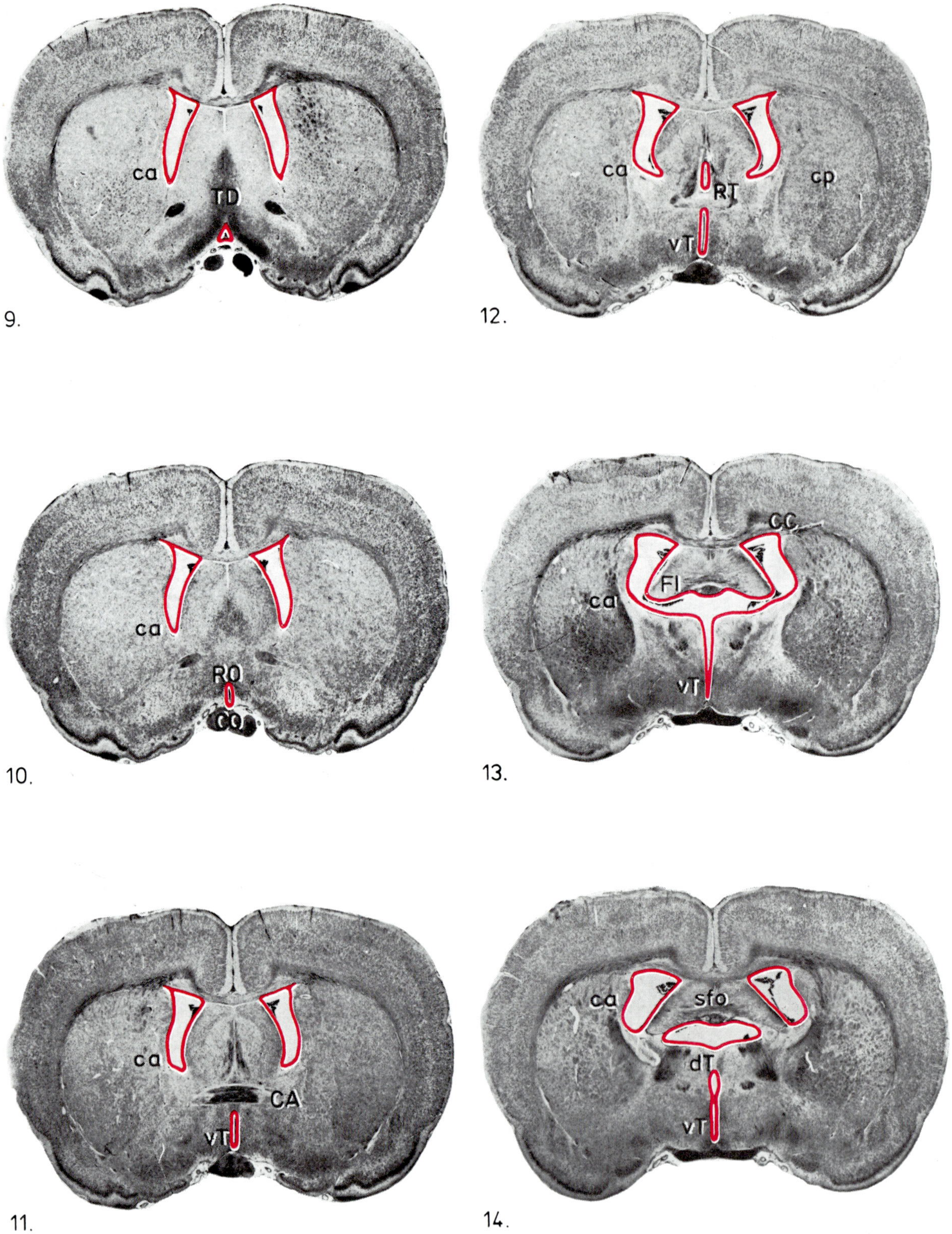

Fig. 2A, sections 9–14

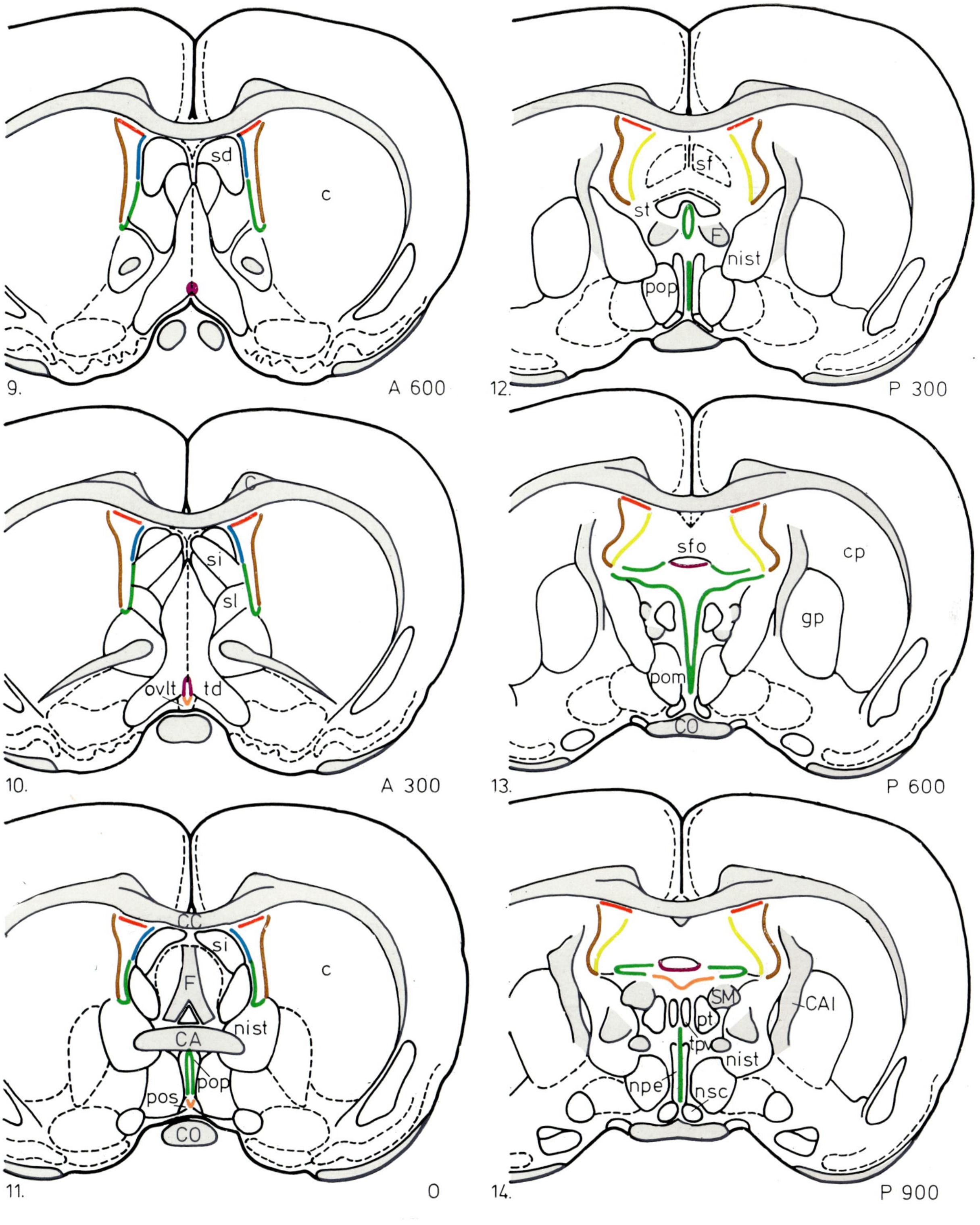

Fig. 2**B**, sections 9–14

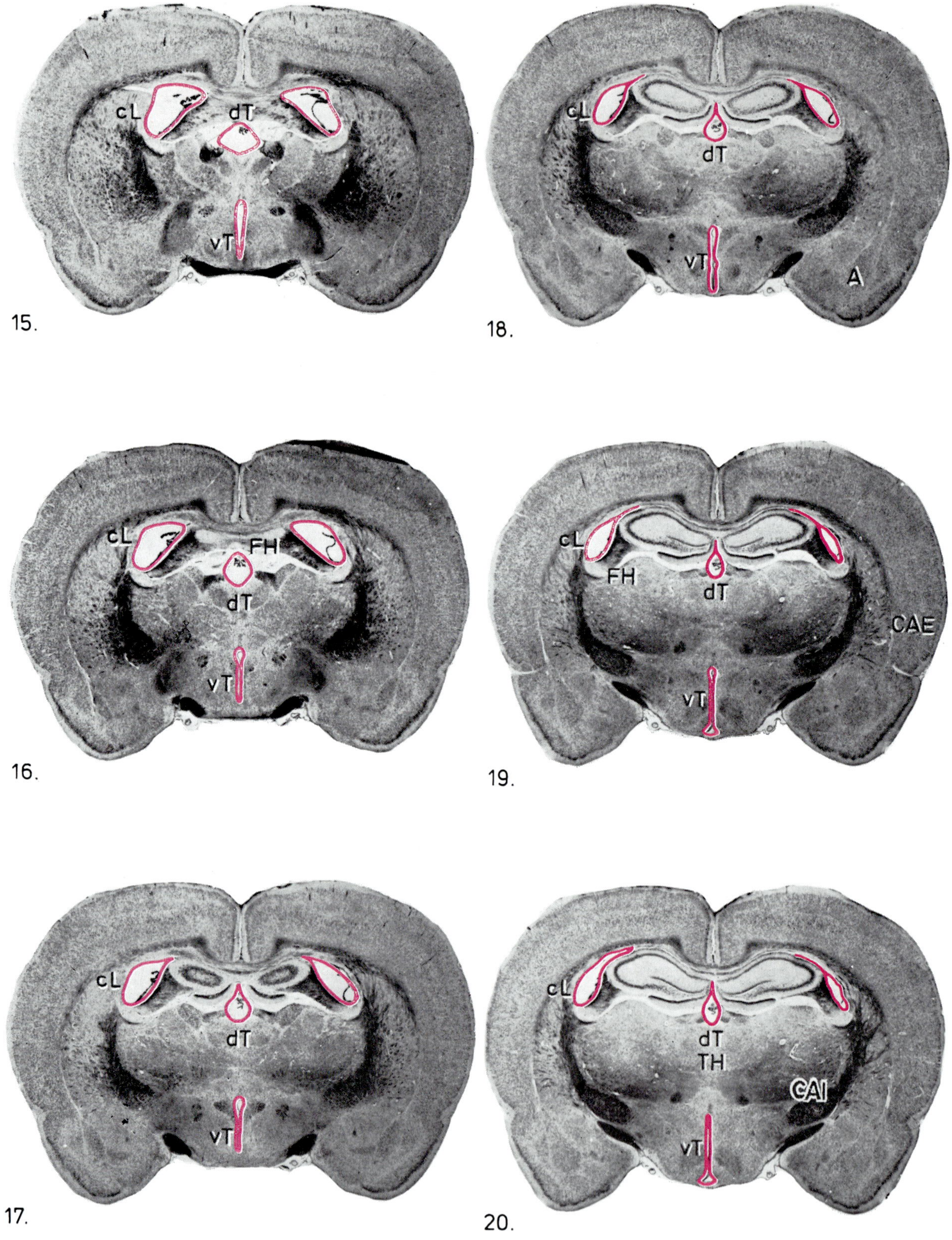

Fig. 2A, sections 15–20

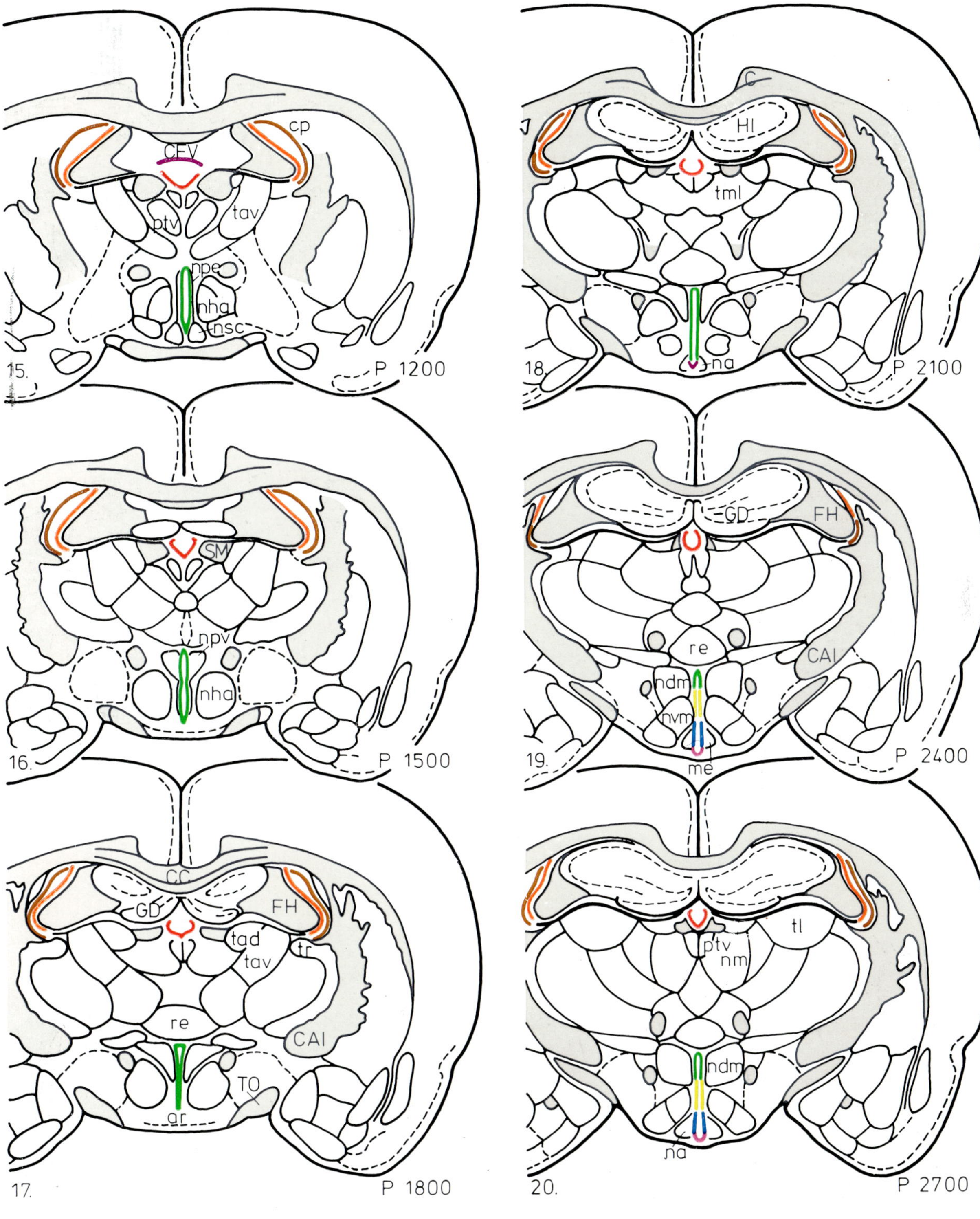

Fig. 2**B**, sections 15–20

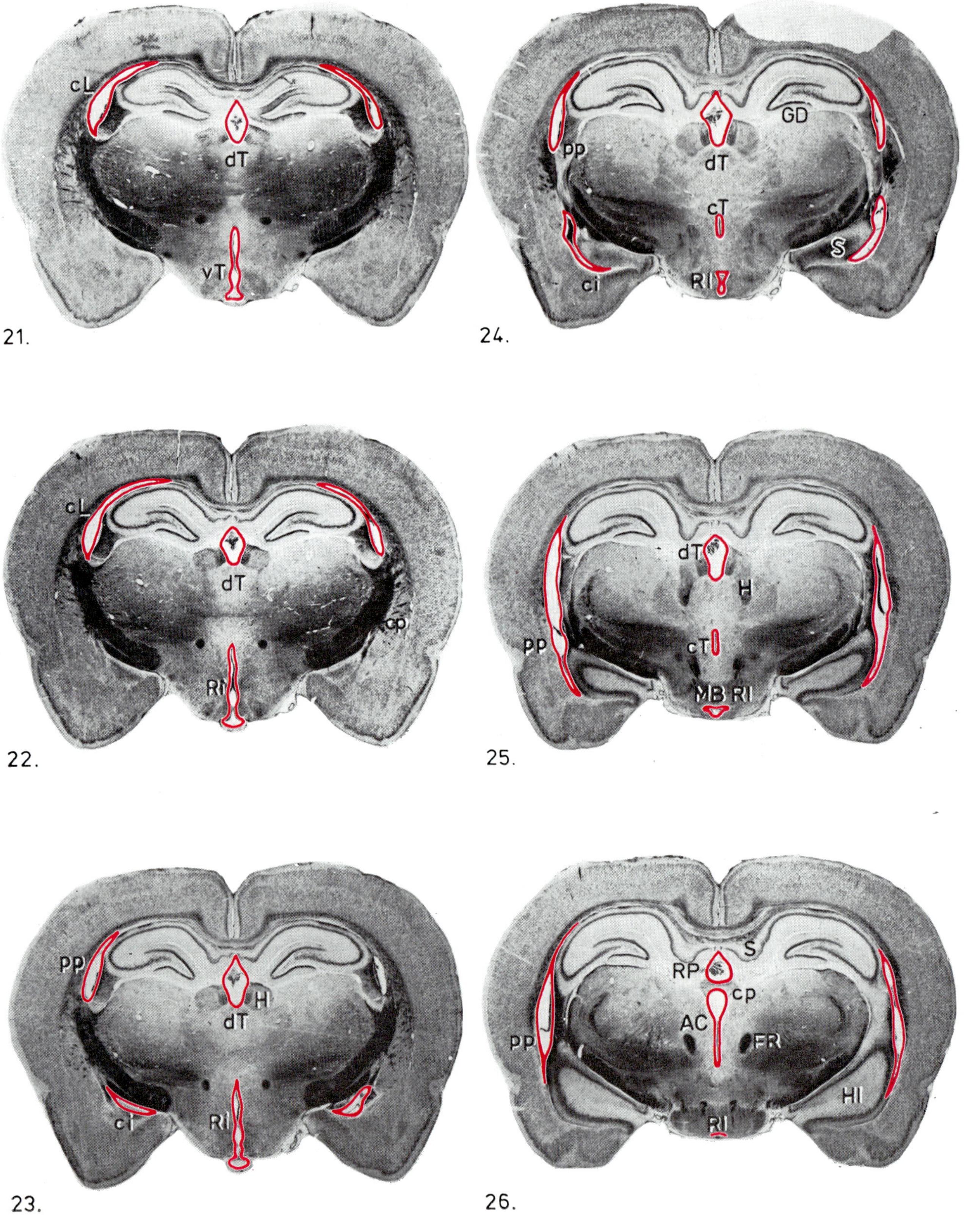

Fig. 2**A**, sections 21–26

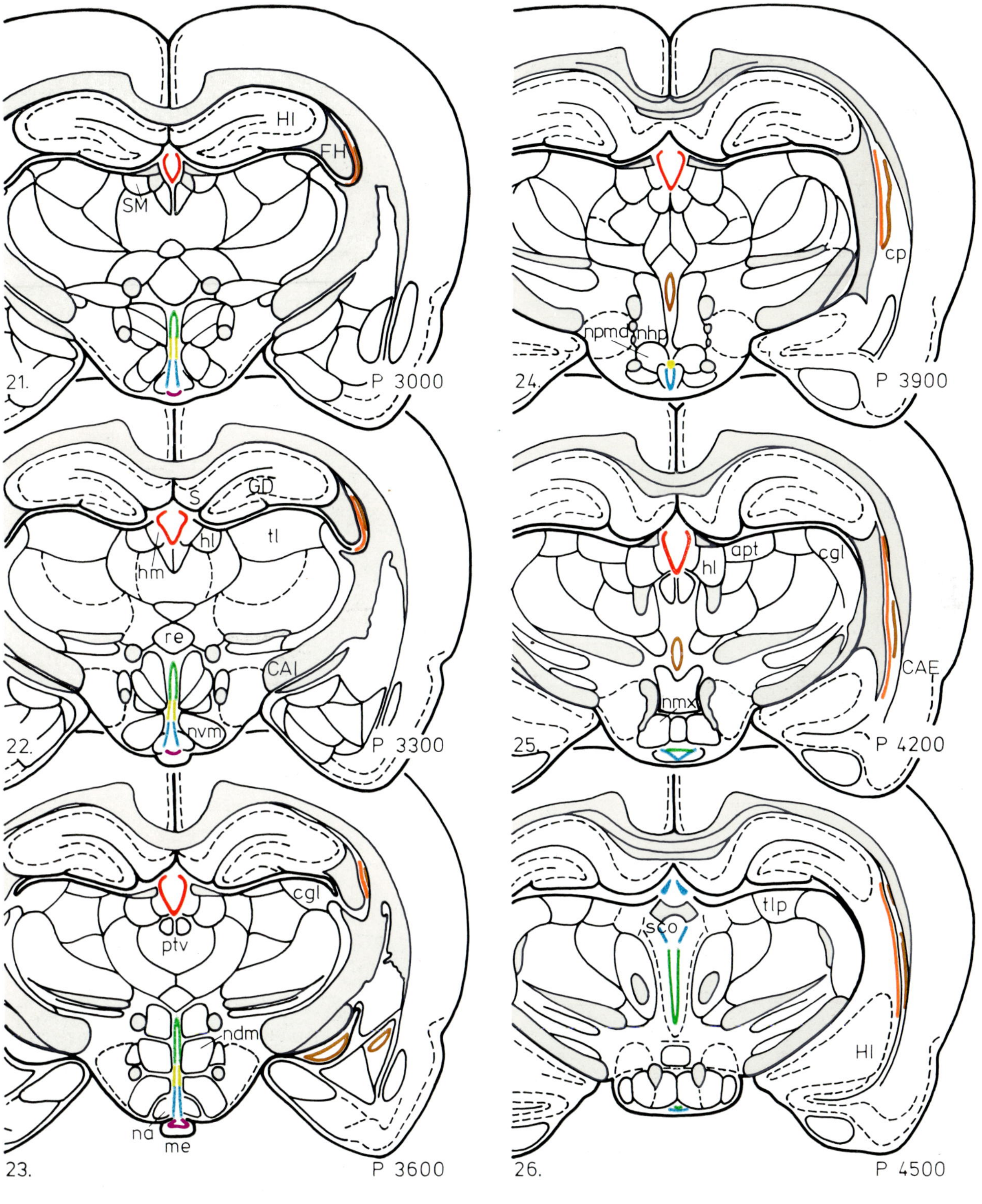

Fig. **2B,** sections 21–26

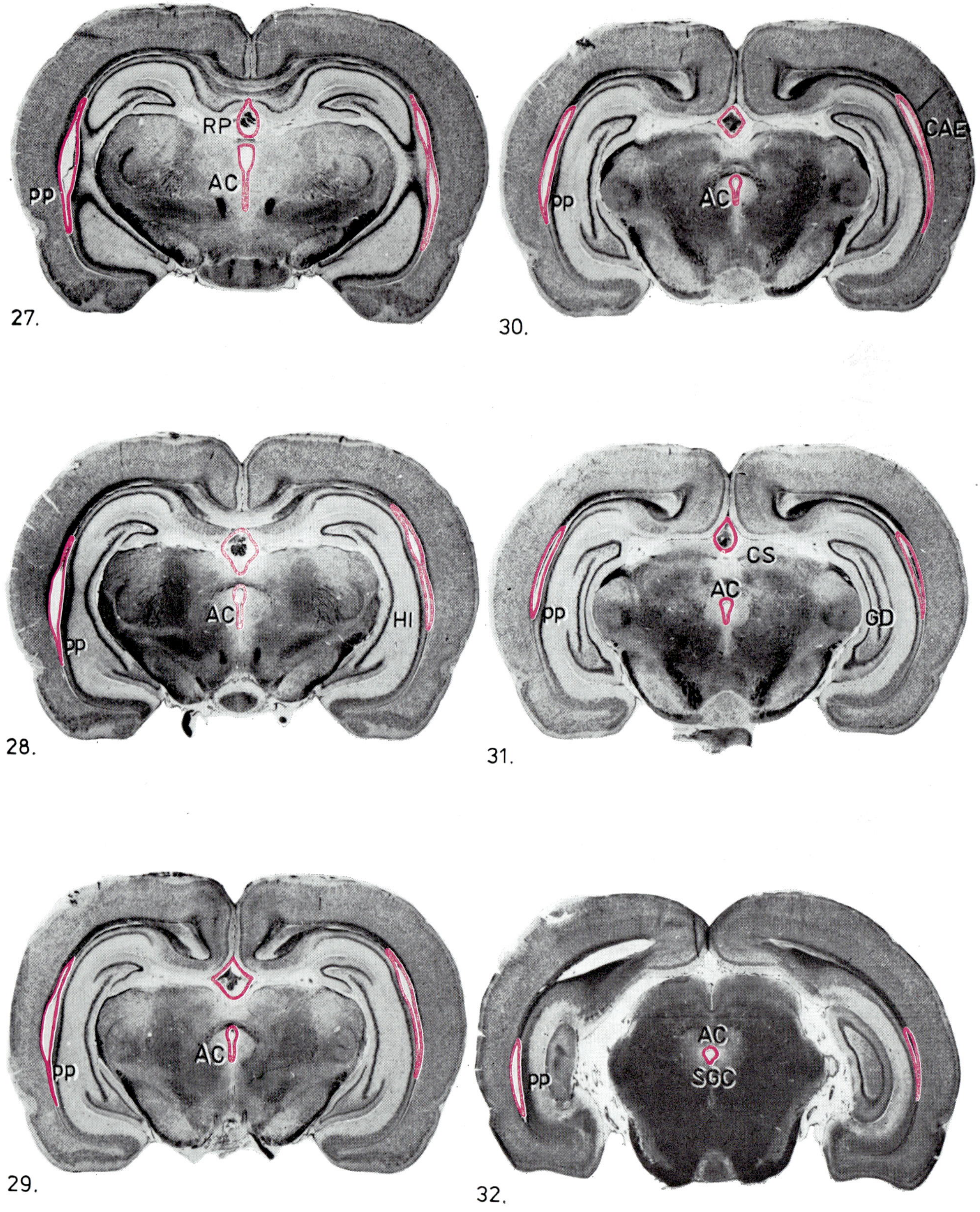

Fig. 2**A**, sections 27–32

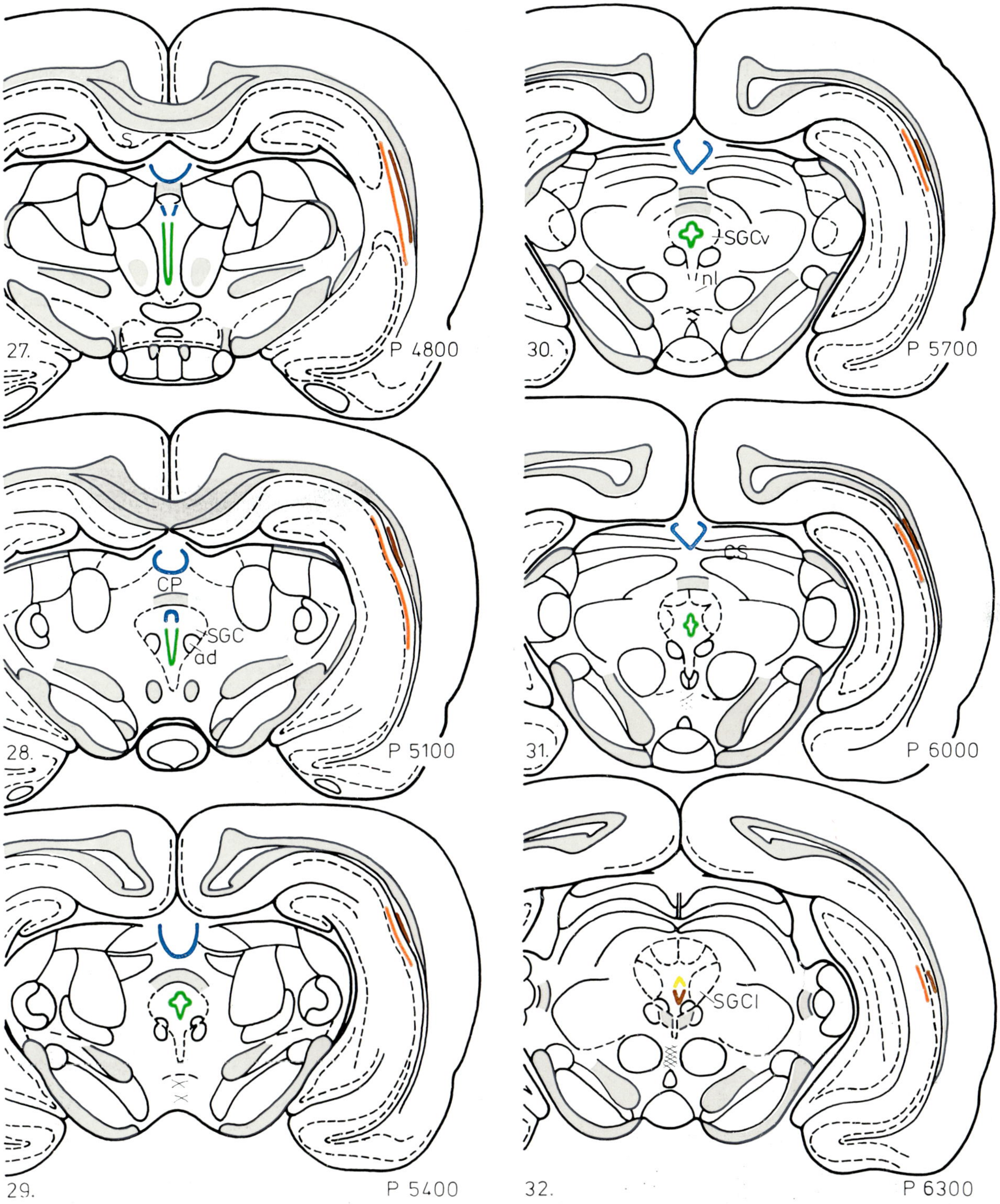

Fig. **2B,** sections 27–32

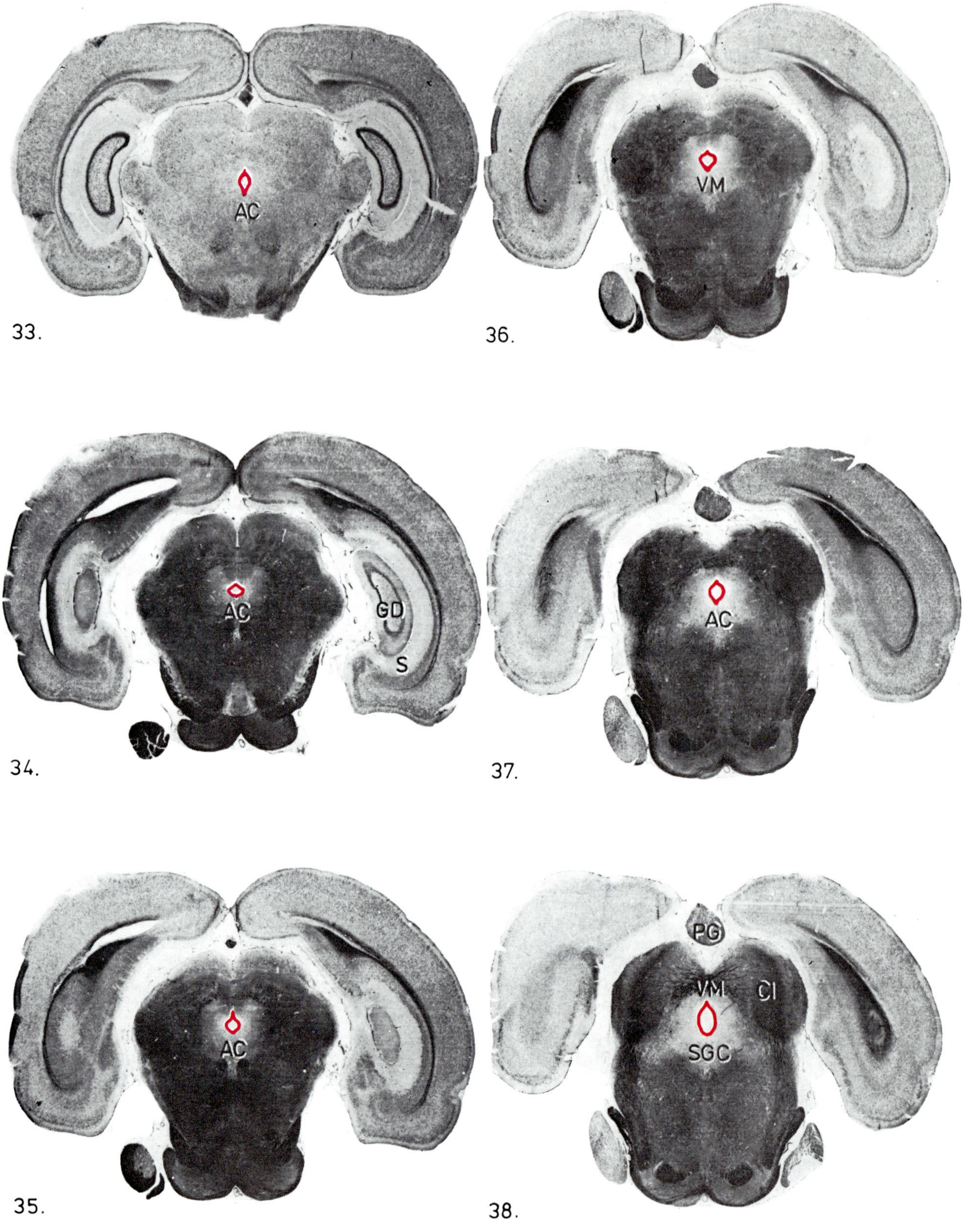

Fig. 2A, sections 33–38

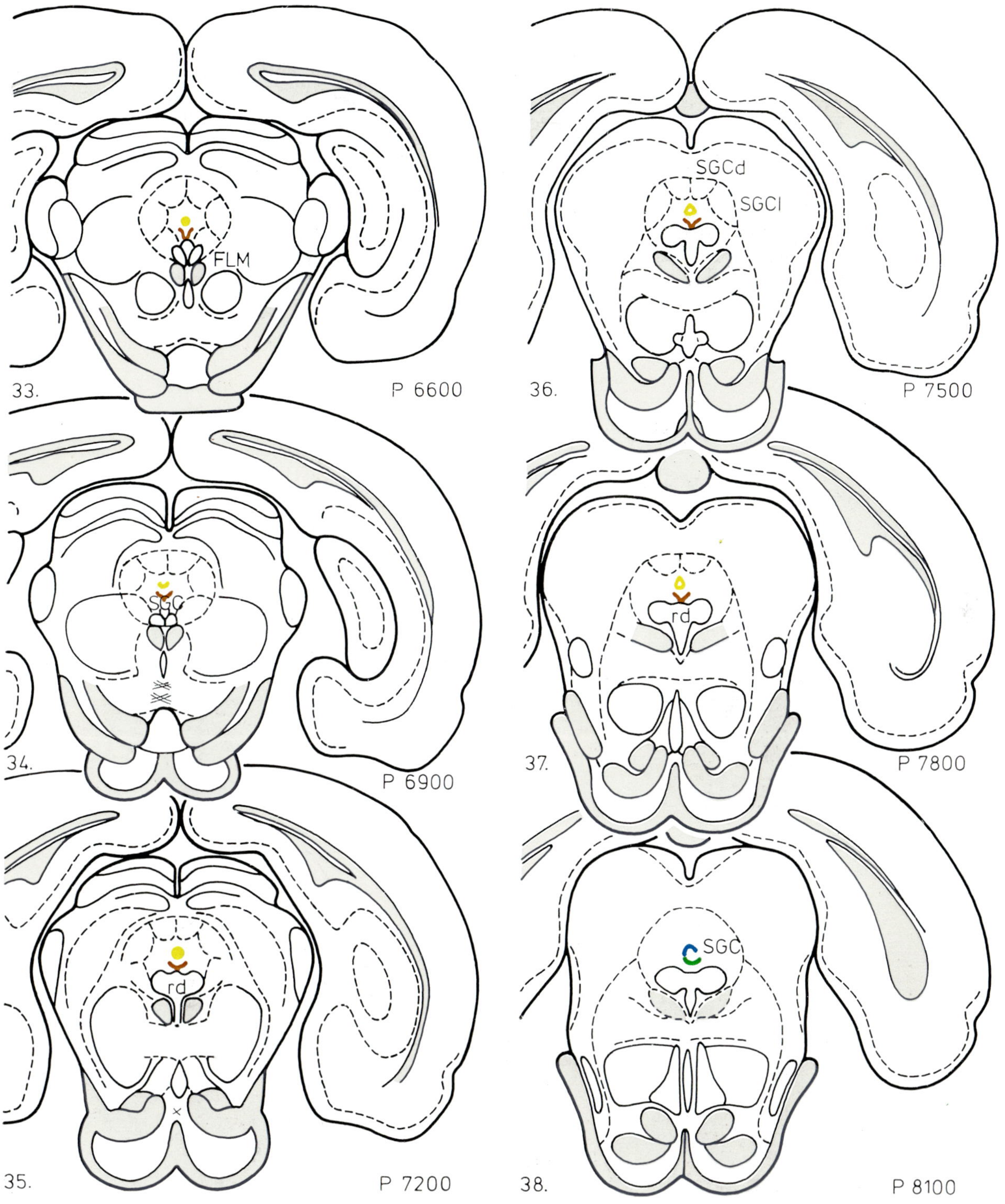

Fig. 2B, sections 33–38

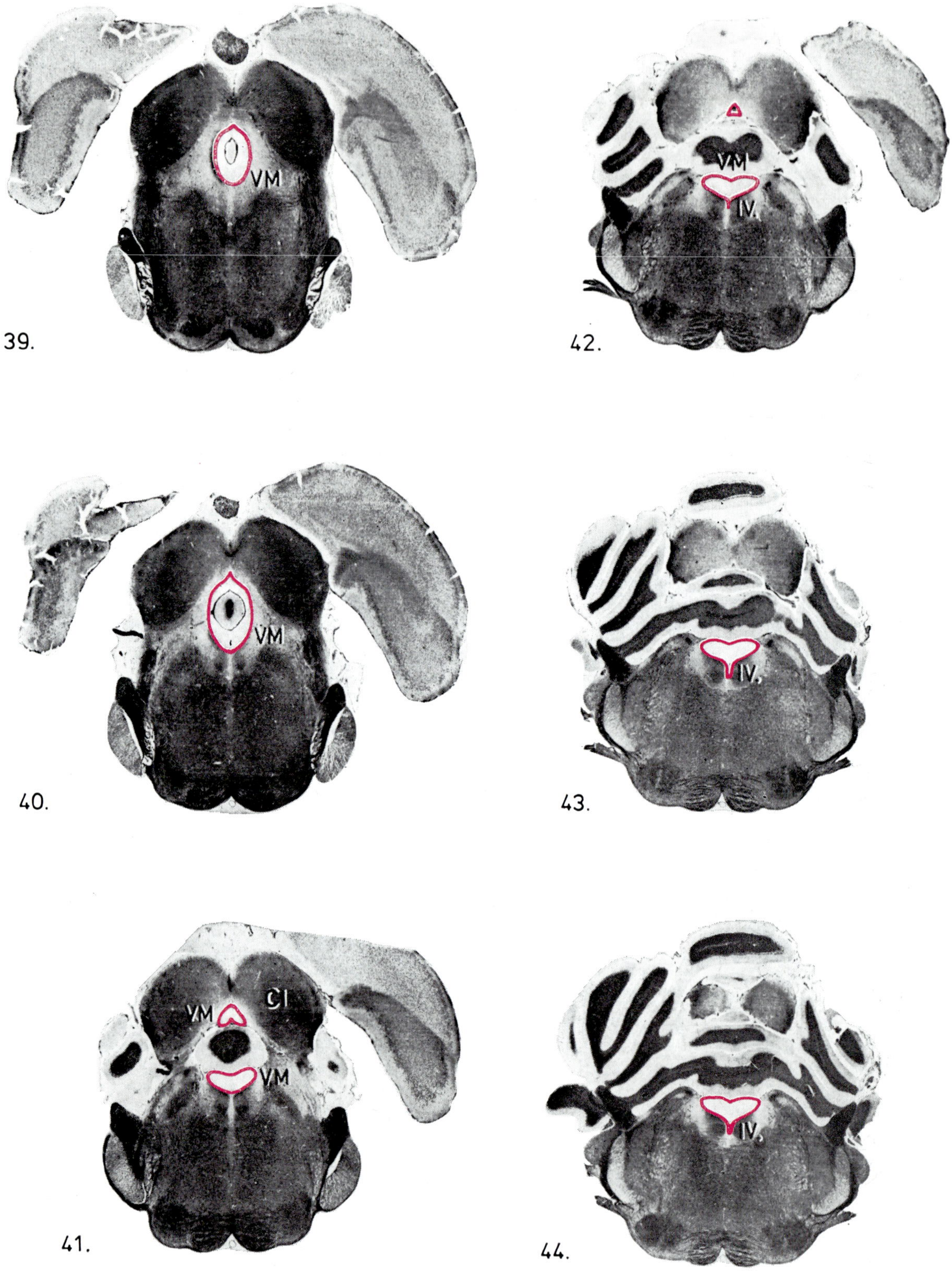

Fig. 2**A**, sections 39–44

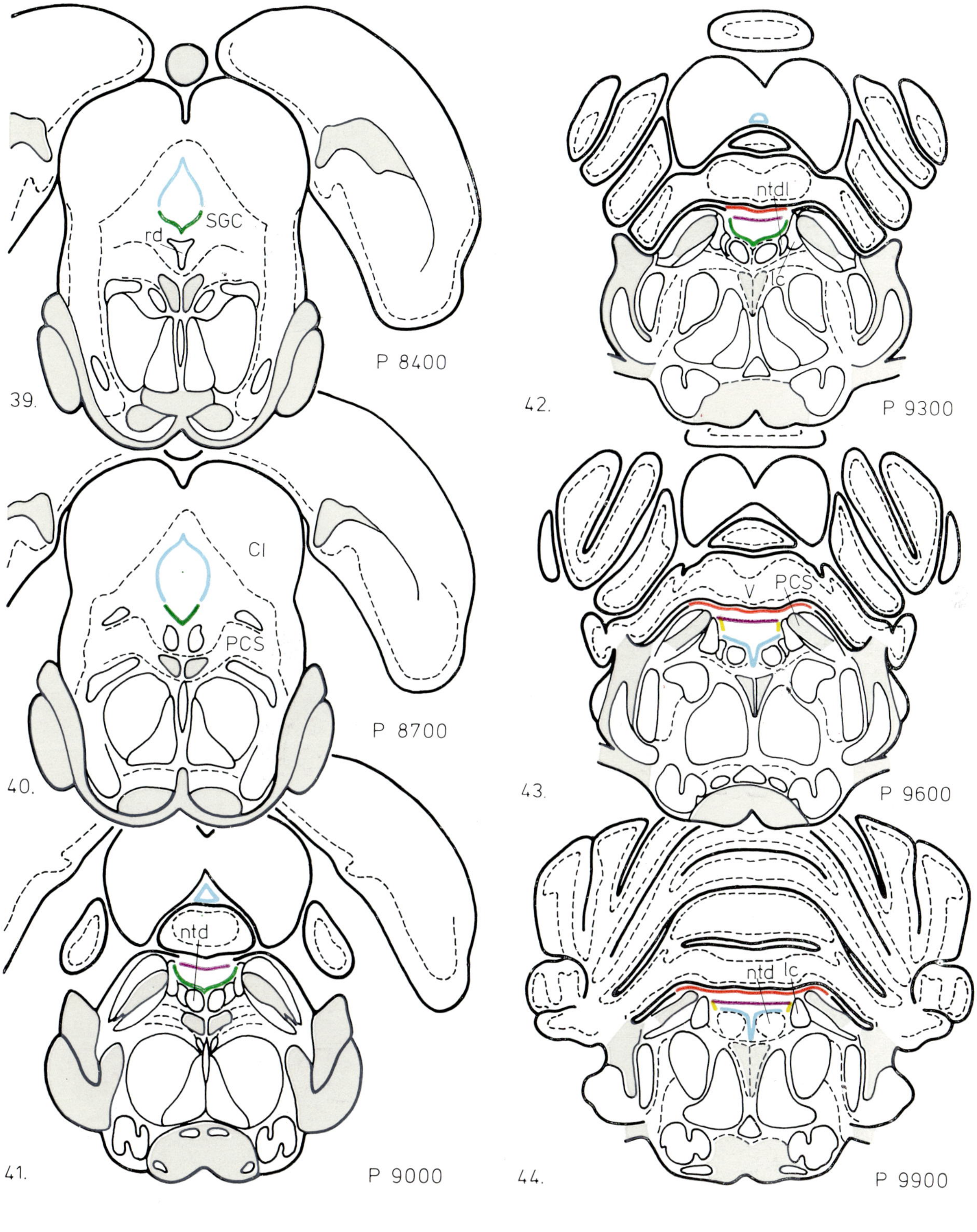

Fig. 2**B**, sections 39–44

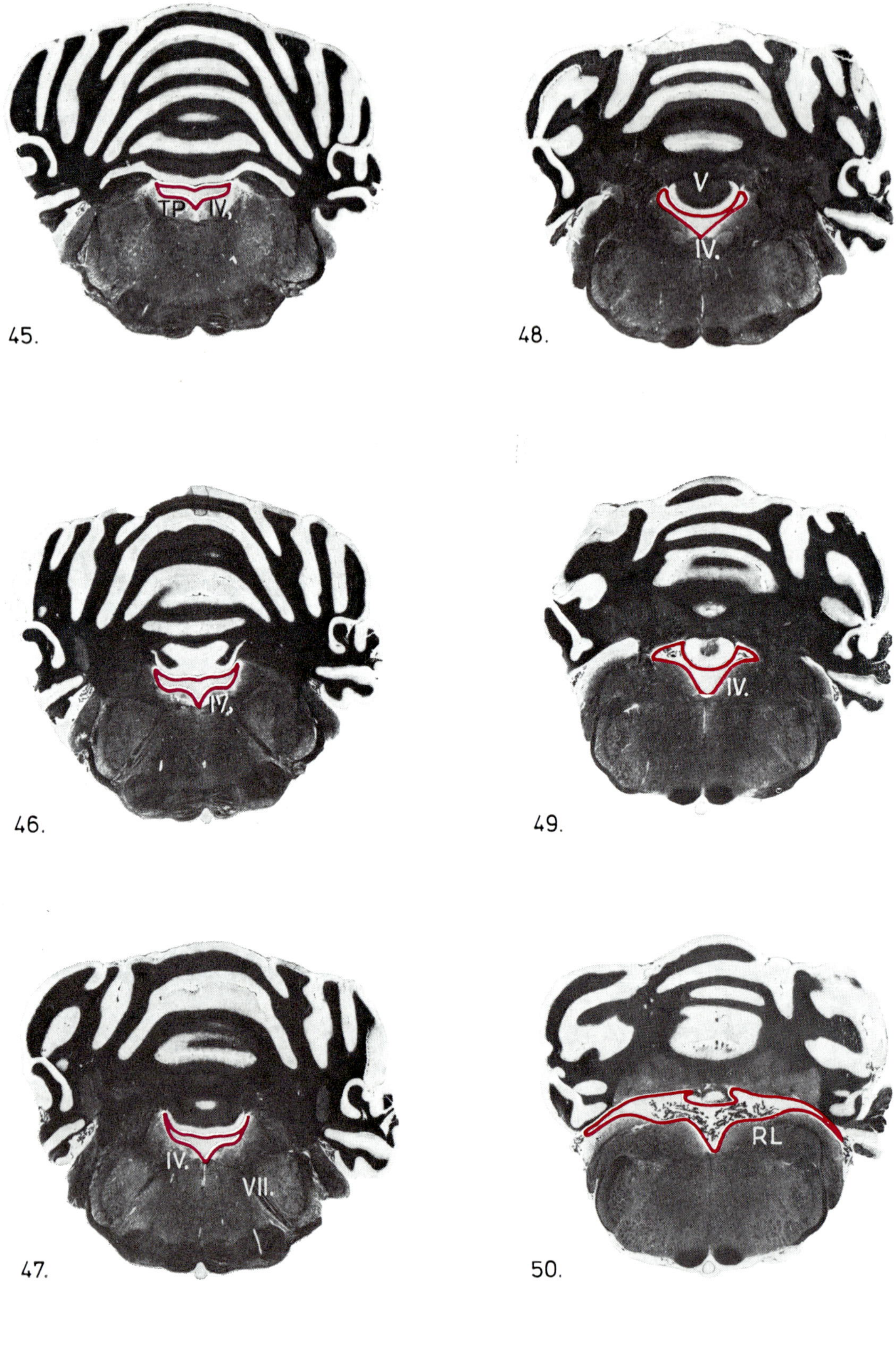

Fig. 2**A**, sections 45–50

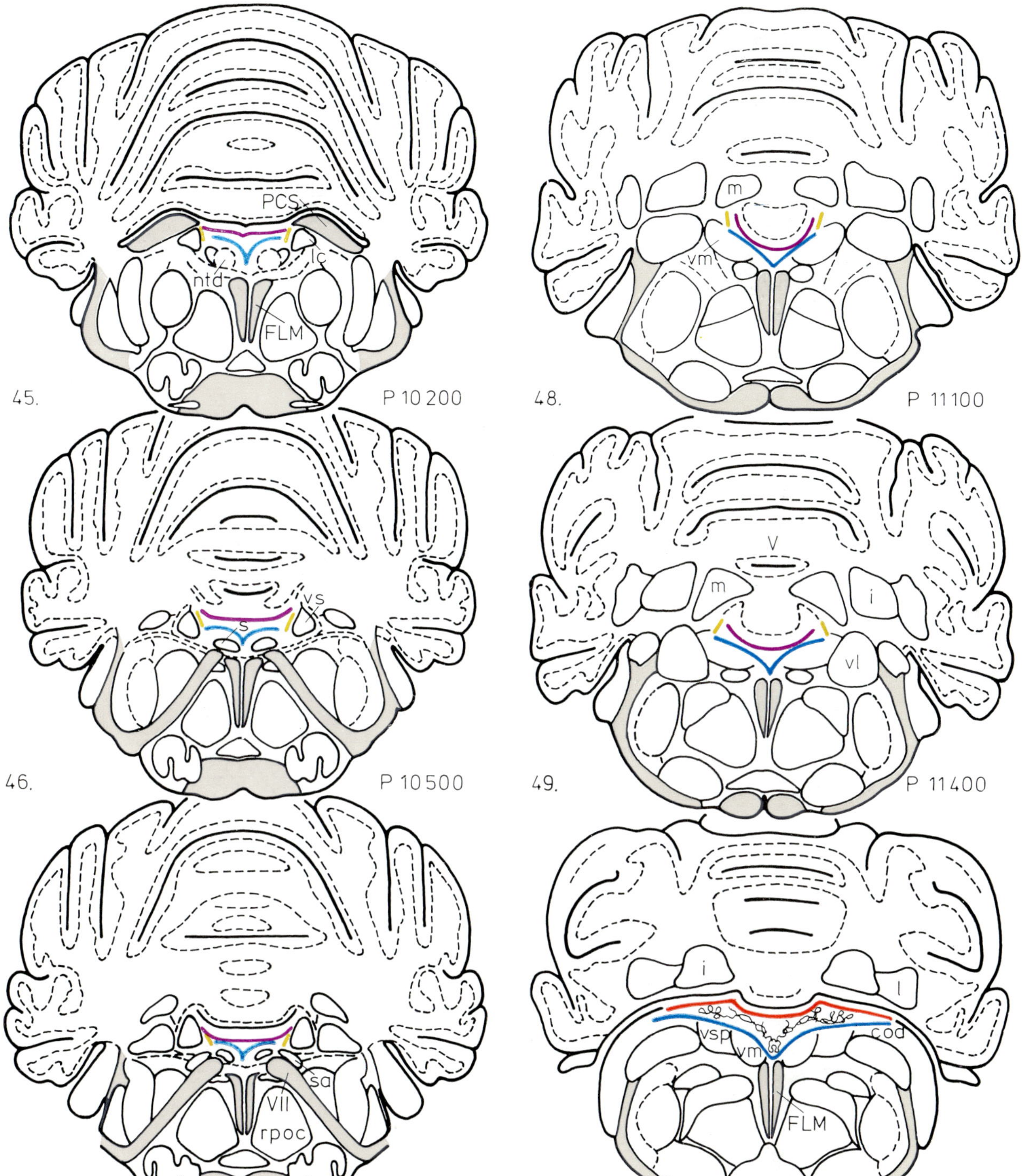

Fig. 2**B**, sections 45–50

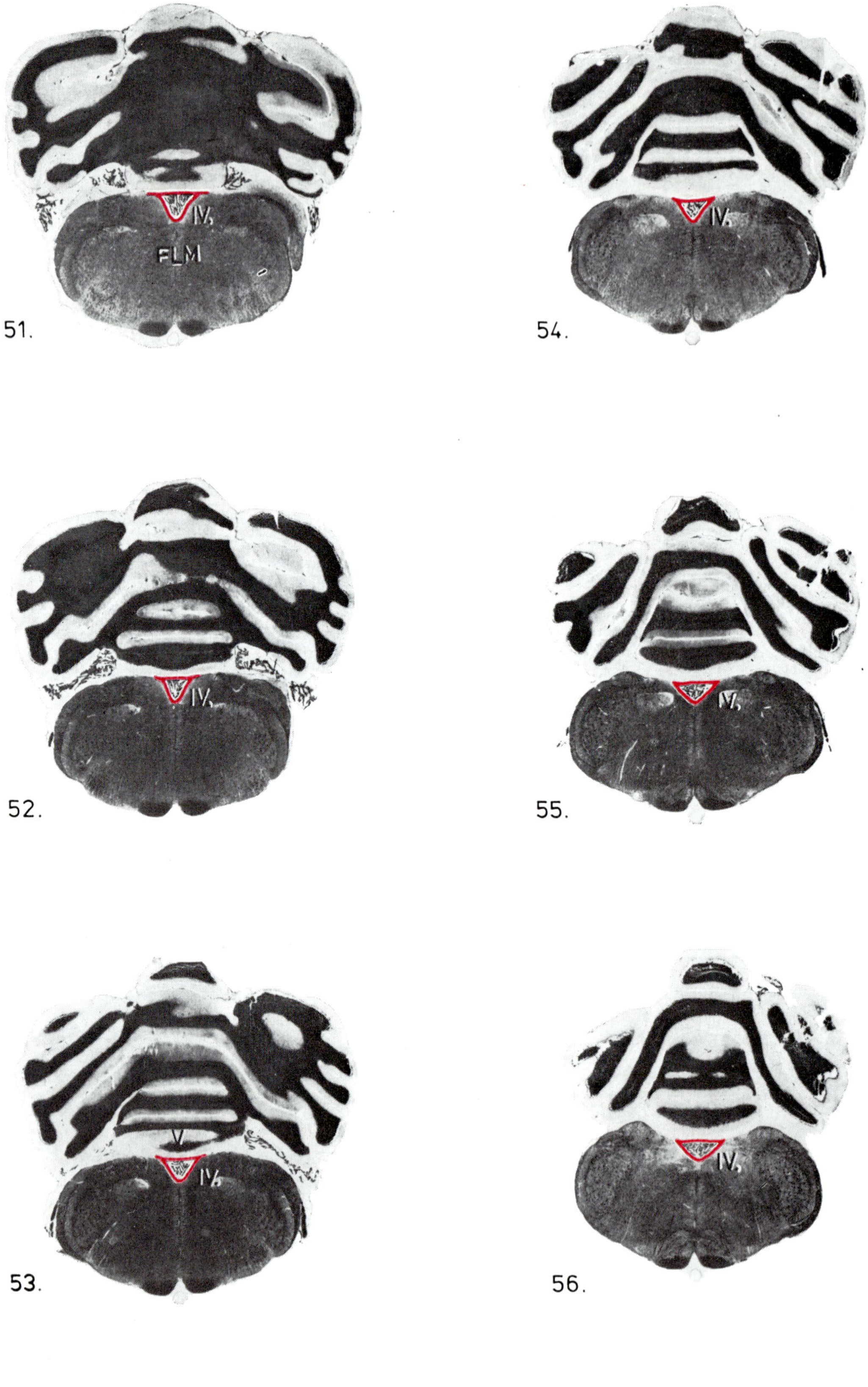

Fig. 2A, sections 51–56

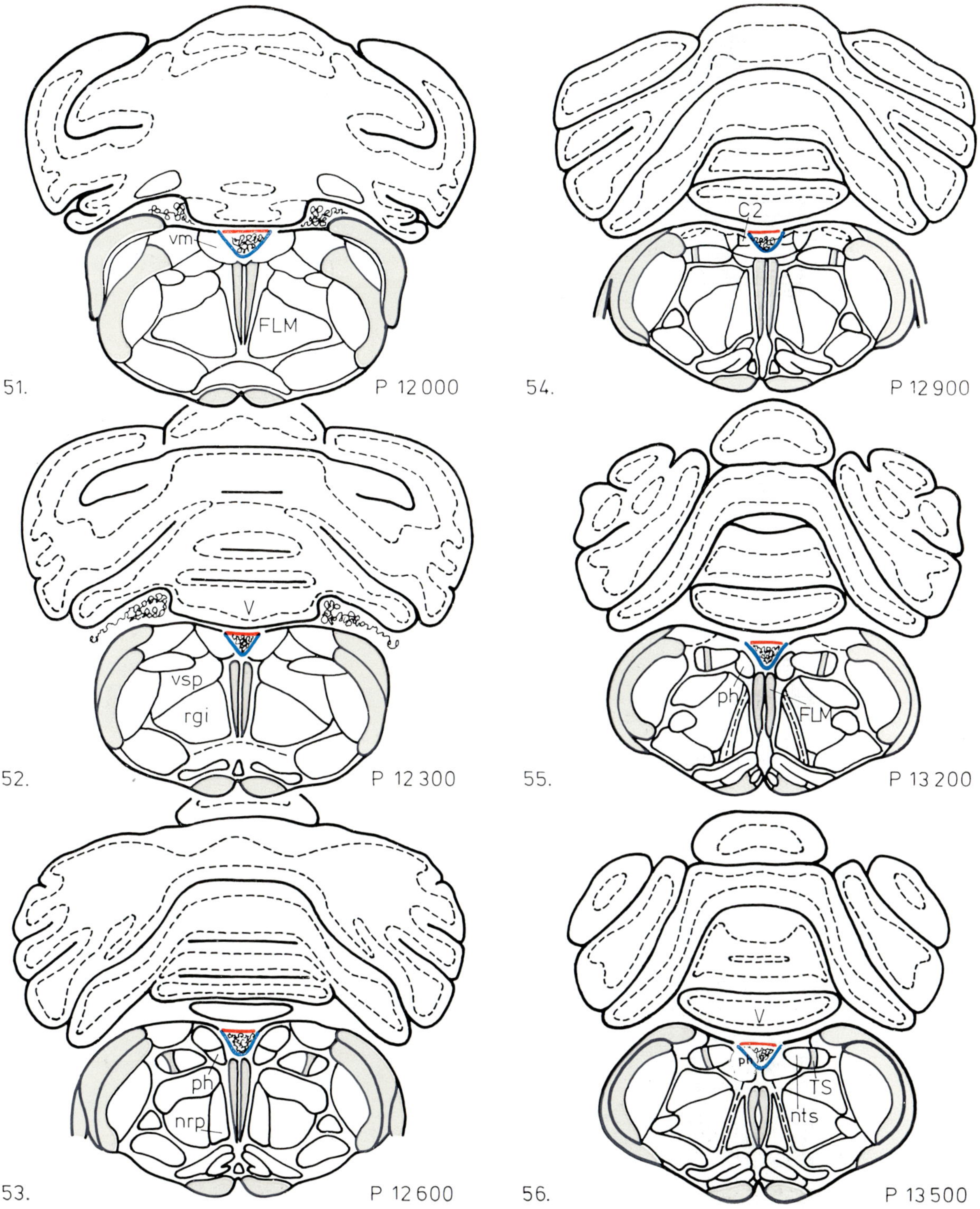

Fig. **2B**, sections 51–56

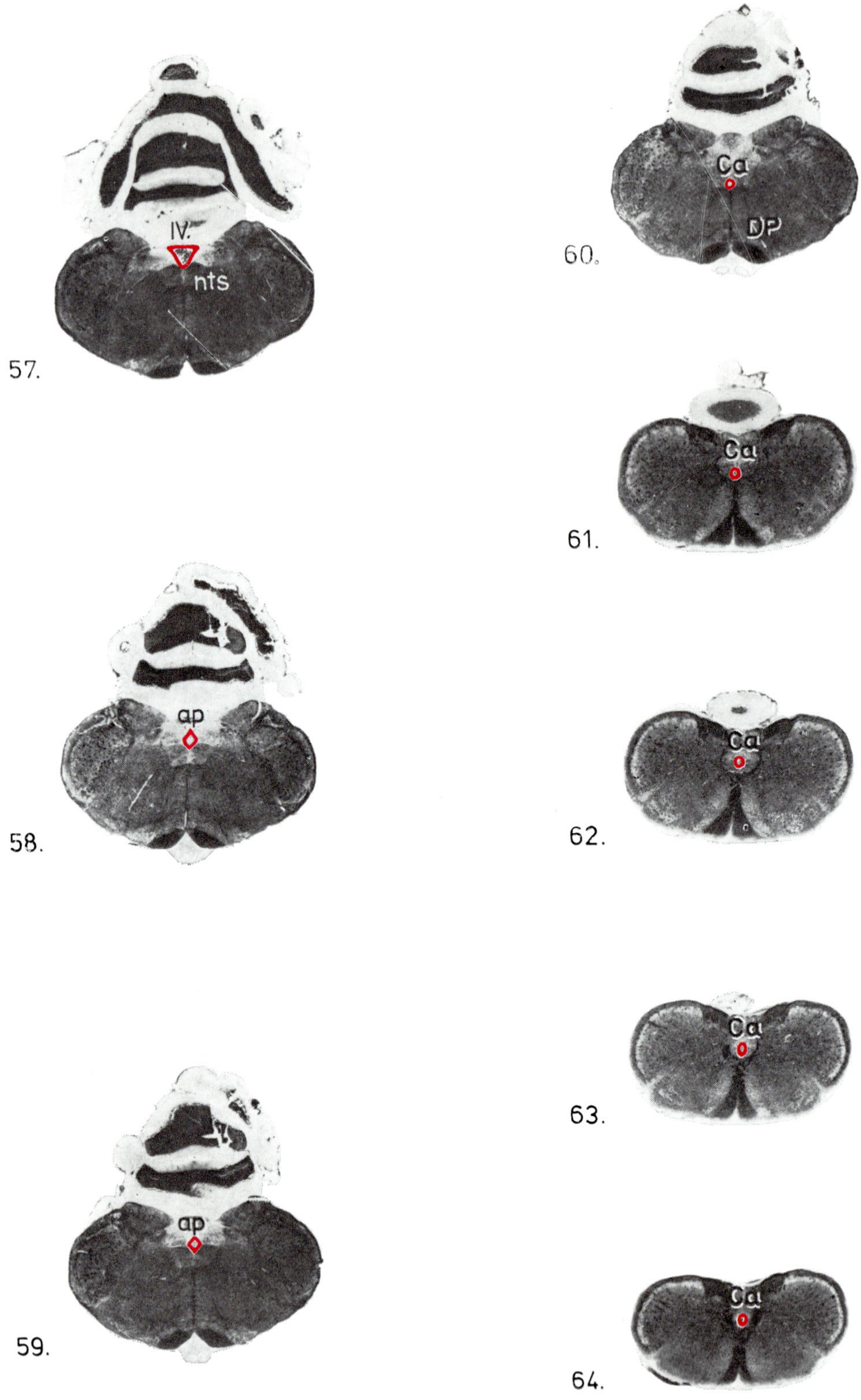

Fig. 2**A**, sections 57–64

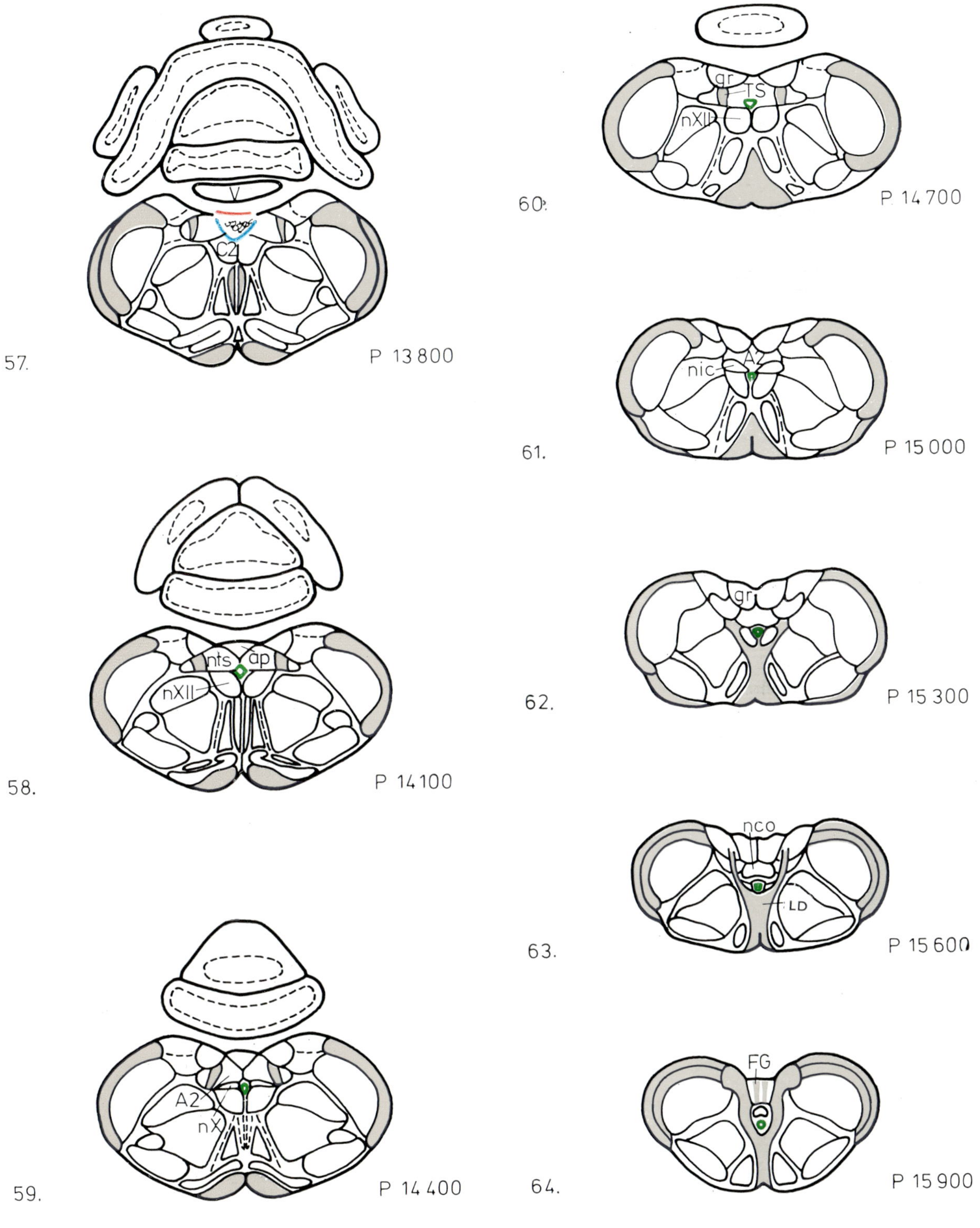

Fig. 2**B,** sections 57–64

LOCALIZATION OF THE VENTRICLES IN SAGITTAL SERIAL SECTIONS OF THE BRAIN

The midline components of the ventricular system can be visualized in midsagittal section (Fig. 3/1). The massa intermedia, surrounded by the dorsal, ventral and connecting parts of the third ventricle with all the recesses, give an impression of the real shape of the third ventricle and allow the size of the compartments of the ventricular system to be compared.

The majority of the ventricles are located in the midline. Laterally the ventricular profiles of the third and fourth ventricles gradually disappear while the lateral ventricles become visible. The anterior horn constitutes the most medial portion of the lateral ventricle (Fig. 3/5). One millimetre from the midline only the lateral ventricle and the lateral recess of the fourth ventricle can be demonstrated in sagittal sections (Fig. 3/7).

LOCALIZATION OF THE VENTRICLES IN HORIZONTAL SERIAL SECTIONS OF THE BRAIN

Mediolateral extensions of the ventricles can be best shown in horizontal sections of the brain (Fig. 4).

In sections cut from the top of the brain the anterior horn and the dorsal tip of the mesencephalic ventricle appear first (Fig. 4/1). They are followed by the whole lateral ventricle except the inferior horn and the dorsal part of the third ventricle in the midline. Topographic relations between the hippocampal and septal fimbria and the lateral ventricle can also be clearly demonstrated. The position of the anterior horn in the mediolateral direction and its communication with the third ventricle through the interventricular foramen are shown in Fig. 4/6.

The ventral portion of the third ventricle, the inferior horn of the lateral ventricle, and the central part of the fourth ventricle can be shown in the same section (Fig. 4/8). The deeper sections are less informative; the section profiles of the ventral portion of the third ventricle and the central canal can be observed as a thin line, or a small oval territory in the midline.

⟶

Fig. 3. Sagittal serial sections (1–8) of the rat brain. Luxol fast blue cresyl violet staining. 10 μm thick sections. Distances lateral (L) to the middle (0 mm) are given in mm. Magnification: ×3.1. For abbreviations see list

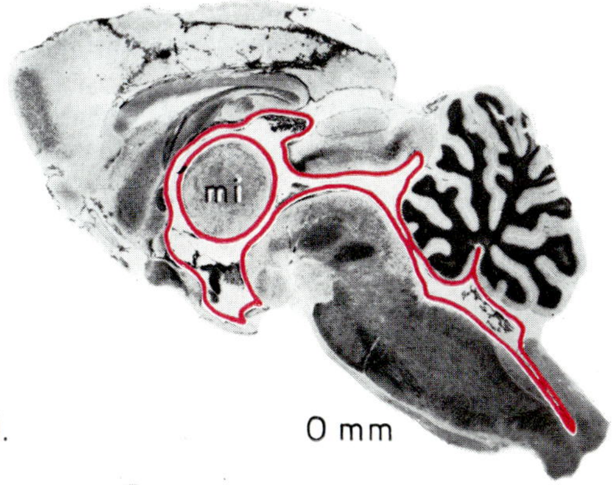

1. 0 mm

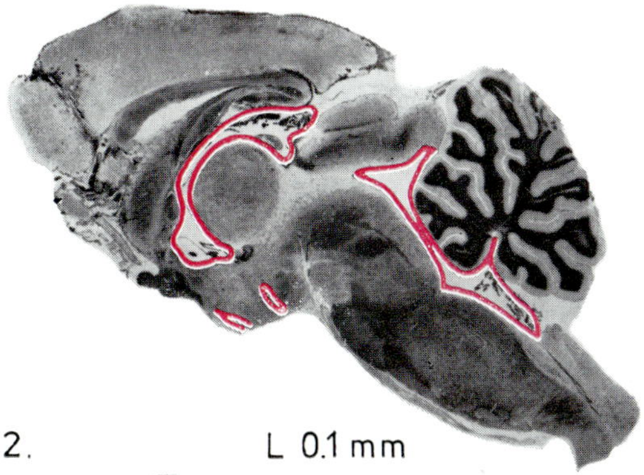

2. L 0.1 mm

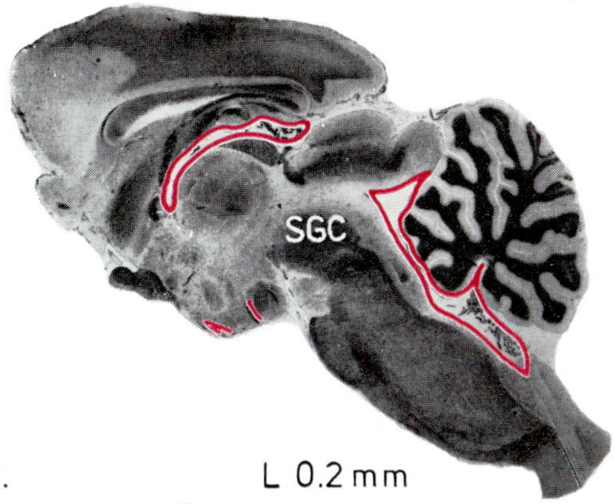

3. L 0.2 mm

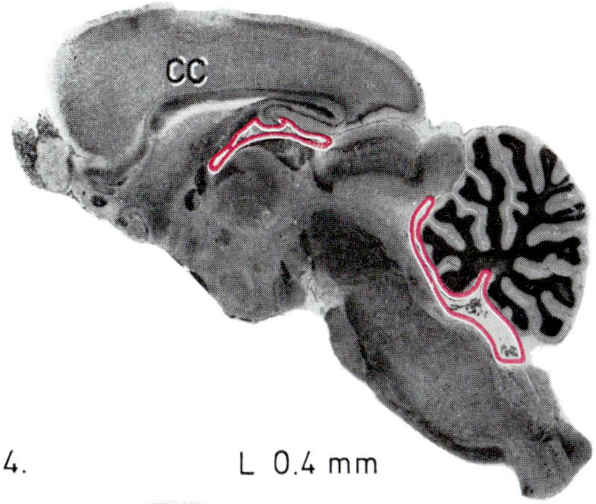

4. L 0.4 mm

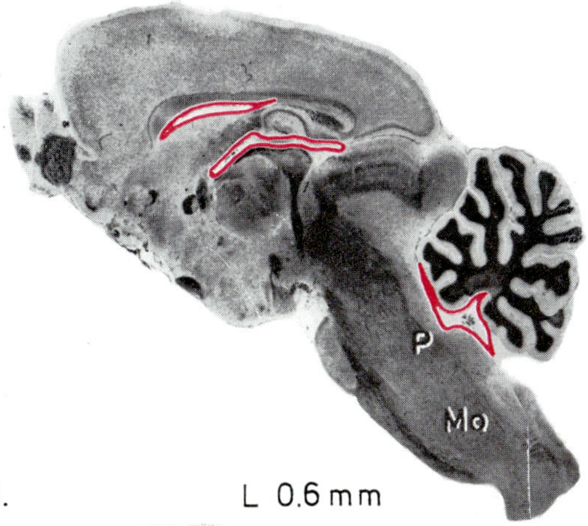

5. L 0.6 mm

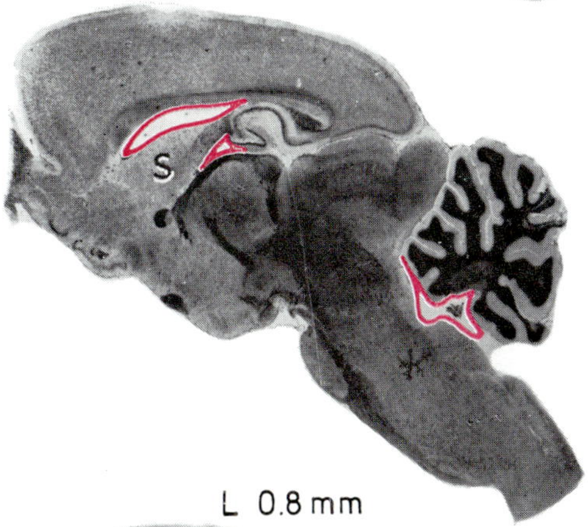

6. L 0.8 mm

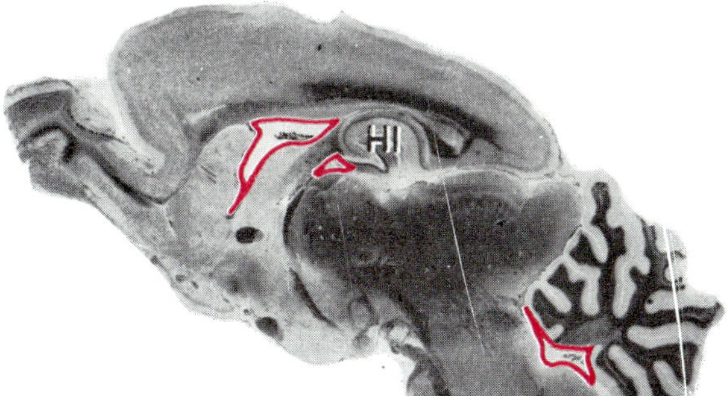

7. L 1.0 mm

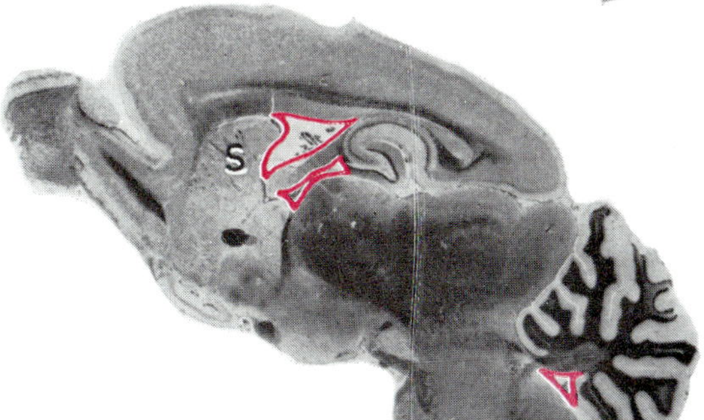

8. L 1.2 mm

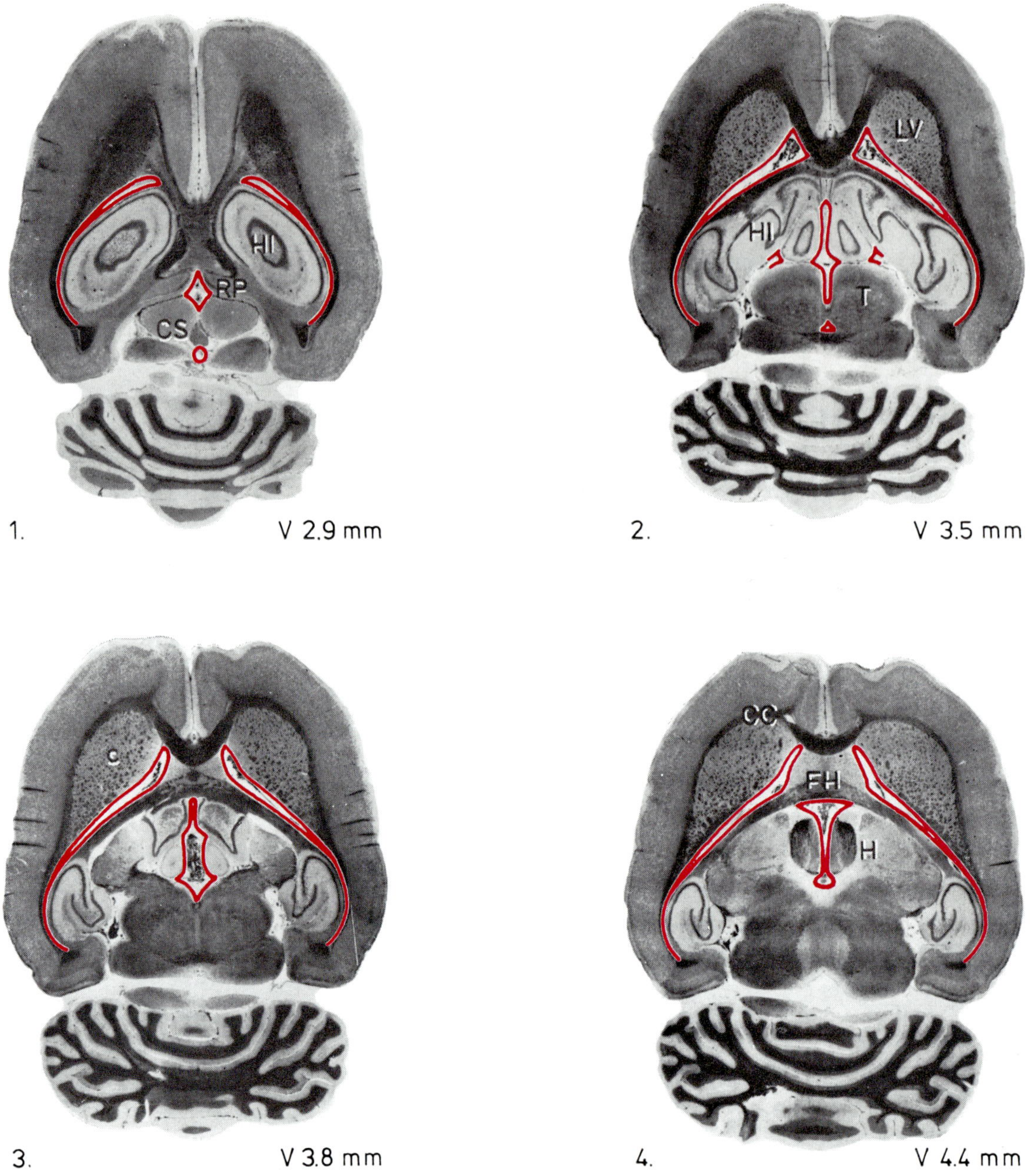

Fig. 4. Horizontal serial sections (1–8) of the rat brain. Luxol fast blue cresyl violet staining, 10 μm thick sections. Distances from the skull (V) are given in mm. Magnification: ×5

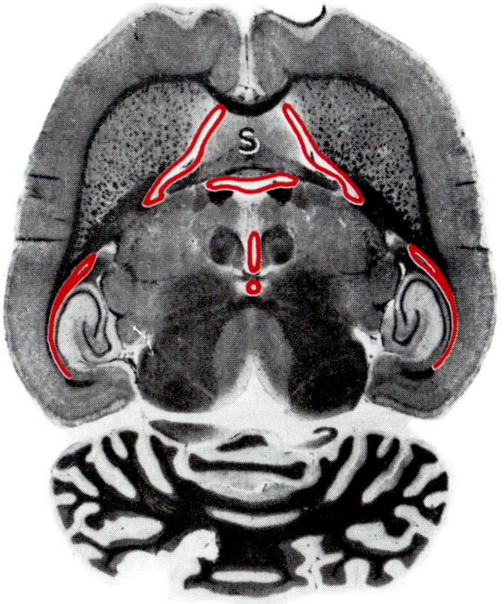

5. V 4.7 mm

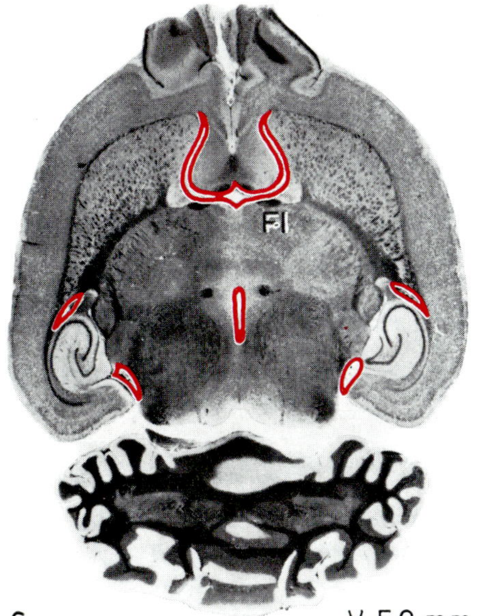

6. V 5.9 mm

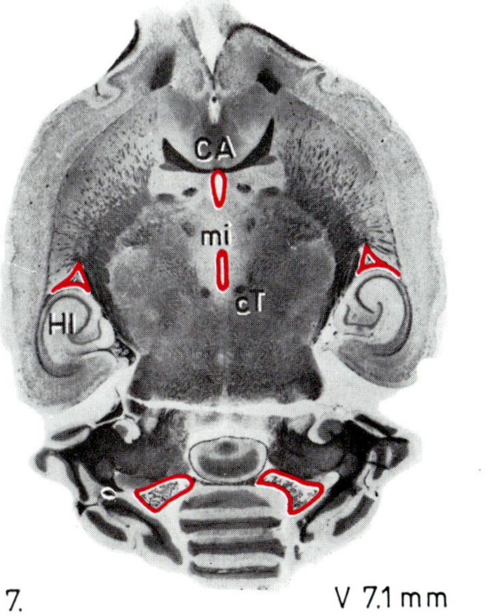

7. V 7.1 mm

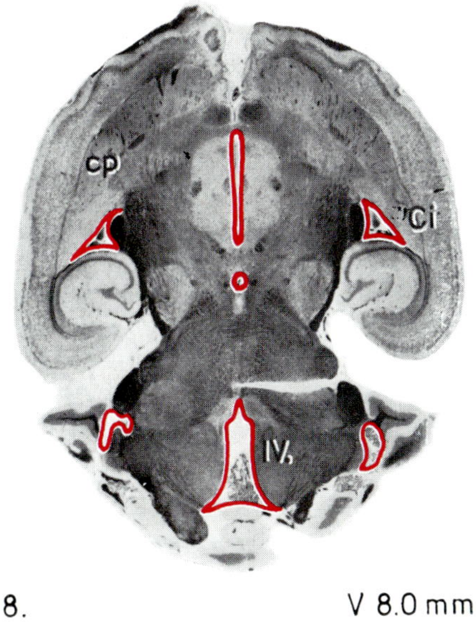

8. V 8.0 mm

EPENDYMAL AREAS OF THE BRAIN VENTRICLES

CURRENT STATE OF KNOWLEDGE ABOUT THE EPENDYMA

A number of morphological studies have been undertaken with respect to the ependymal lining of the wall of the ventricles (for review see Sterba, 1969; Knigge et al., 1972, 1975; Mitro, 1974; Sterba and Bargmann, 1977). They all agree that the ventricular lining is very complex (Peters et al., 1976).

Scanning electron microscopy has revealed a great variability of the surface of the ventricular wall (Bruni et al., 1972; Noack et al., 1972; Allen and Low, 1973; Coates, 1977; Lindberg et al., 1977; Shioda et al., 1977). This technique allows a large surface area to be examined with a rather high resolution. Scott et al. (1974a) gave an account of their scanning electron microscopic investigations of the whole ventricle system. The dorsal wall of the third ventricle is characterized by a profusion of ciliated ependymal cells. Underlying microvilli are usually hidden by a dense overlying mat of cilia. Ependymal cilia generally appear uniform. At about the middle third of the cerebral ventricle, a gradual decrease in the number of cilia and a modification in surface structures become evident. The ventral third and the floor of the third ventricle are characterized by the virtual absence of cilia. In their place numerous small villiform processes could be observed. A dominant feature of the floor and infundibular recess are the large cytoplasmic protrusions of the apical surface of the ependymal cells. Unusual "beaded" or "club-like" processes are also present. Curious stellate "free cells" (neuron-like in appearance) are frequently observed on the surface of the ventricular lumina. The presence of axonal processes and/or terminals was demonstrated. Occasionally it is possible to visualize vessel-like structures that appear to traverse the infundibular recess. In the caudal wall of the mamillary recess strongly ciliated cells were seen. The luminar surface of the subfornical organ is covered by numerous long cilia and delicate microvilli. The density of the cilia here is not comparable to that of the dorsal thalamic wall of the third ventricle. At the level of the pineal recess there is an abrupt transition between the ciliated and non-ciliated surfaces. There is a marked loss of cilia; they are replaced by microvilli covering the apical surface of the ependymal cells. The floor of the suprahabenular and intercommissural recesses of the rat is covered by non-ciliated ependyma (Ribas, 1977b). The cilia are predominant surface organelles of the lining ependymal cells in the lateral ventricle, however, microvilli are also present in a limited number. Cilia are the densest over the areas of the white matter, whereas over the gray matter they occur in groups. According to Peters (1974), each ependymal cell has a taft of 20 to 30 cilia in addition to many small and thin villous processes. Brawer et al. (1974) examined the wall of the third ventricle of rats at each stage of the oestrous cycle and found irregularities at apical membranes of tanycytes. Tanycytes were shown to be devoid of microvilli and of other surface irregularities. The supraependymal cells of the third ventricle show great polymorphism. In addition to bipolar, multipolar and stellate or spider-like cells, transitional forms between these types can be seen (Mesters and Breipohl, 1976). The regional variation in the distribution of supraependymal nerves of the rat epithalamus was observed by Ribas (1977b). The ventricular surfaces of the choroid plexus cells are domed and have a tangle of thin processes resembling microvilli. No cilia are present. There are a number of free cells overlying the choroid plexus. These cells emit many fine processes (Peters, 1974). In the third ventricle, such as microvilli, craters and areas of discontinuous lining were described as surface specializations of the infundibular recess (Martínez Martínez and de Weerd, 1977). The rostral half of the floor of the third ventricle was densely ciliated, while the caudal portion possessed a few cilia. The ciliated cells ended in an abrupt transition zone located about halfway along the floor (Paul et al., 1977). Jennes et al. (1977) investigated the formation of the Reissner's fibres in the rat. Intraventricular macrophages were observed with the scanning electron microscope on the choroid plexus in the developing rat brain (Chamberlain, 1974).

Bleier (1975) studied the ependymal surface of the hypothalamic third ventricle of adult female and male mice. Cells have no cilia, microvilli or membrane folds, though they may have small bleb-like protrusions. Some cells and cilia are covered with clusters of secretory-like material. Cilia are densely packed in the medial part and are sparse in the lateral part of the fourth ventricular floor of the mouse (Yamadori and Yagihashi, 1975). The direction of the ciliary flutter was determined by scanning electron microscopy in the fourth ventricle of the mouse. It is either toward the central canal or toward the lateral apertures of the fourth ventricle. These observations suggest that the ciliary flutter regulates the direction of the cerebrospinal fluid streaming in the ventricle (Yamadori, 1977).

The surface of the ependymal cells of the cat were investigated with scanning electron microscopy by Clementi and Marini (1972). The ventricle walls are entirely covered with numerous long cilia. Regarding their number and position, no regional differences are encountered.

Transient morphological changes in the ependymal surface of the ventrolateral wall and floor of the third ventricle of the rabbit were observed by Bruni et al. (1977) 5 min after LH–RH injection. The numerous irregular microvillous eruptions of the apical membrane may reflect in their opinion structural modifications associated with the absorption of the hormone across the ependymal–CSF interface.

The development of the ependyma in the third ventricle of the hamster was investigated by Hannah and Geber (1977). Ependymal cells exhibit large apical protrusions which contain ribosomes and cytoplasmic matrix. The transitory specialization of the ependyma is proposed to represent either a neurosecretory or a mechanical folding mechanism.

Dalen et al. (1971) studied cilia on cultured ependymal cells in different developmental stages by scanning electron microscope.

Coates (1973) characterized the population of supraependymal cells in Rhesus monkeys of both sexes. Round cell bodies possess multiple branching processes extending from the cell body over underlying ependymal surfaces. Such processes have varicosities along their course. Small spherules are present on cell bodies and on processes. Some processes apparently disappear through holes and crevasses in the ependyma.

Jacobs and Monroe (1977) investigated by scanning electron microscopy the ventricular system of the female armadillo.

The technique of freeze-fracture was also employed to study the ependymal lining of the brain ventricles. The analysis of replicas obtained from freeze-fractured material provides an additional tool to correlate membrane features (macromolecular composition of membrane) as seen with this technique and with conventional electron microscopy.

Membrane specializations corresponding to gap junctions and attachment plates were found on ependymal cells. The membranes of the subependymal cells were characterized by the paucity of intramembraneous particles (Privat, 1977).

The freeze-etching technique was employed for the in vivo and in vitro study of the development and differentiation of the neuroepithelial junctions in the plexus epithelium. The in vivo formation begins with the appearance of amorphous patches of the particles; this is followed by the formation of small particular rows and polygonally ordered particle assemblies. The in vitro reconstruction of the zonula occludens proceeds from remnants of the former zonula occludens. On the 7th day, the cultured choroid epithelial cells display well-differentiated junctional complexes consisting of continuous zonulae occludentes and integrated gap junctions (Dermietzel et al., 1977).

Freeze-etched preparations of the ventricular surfaces of ependymal cells clearly reveal the presence of pinocytotic vesicular openings into the third ventricle and large vacuoles formed by broad cell protrusions. Gap junctions between the adjacent ependymal cells were detected near the median eminence in the mouse but not in the frog (Nakai et al., 1977b).

The ependymal cells develop from the cells of the "ventricular zone" (Boulder Committee, 1970) from which nerve and glial cells develop. Accordingly, the ependyma represents a regular component of the nervous tissue (Schachenmayr, 1967; Schiebler and Mitro, 1969; Westergaard, 1969; Kirsche, 1970; Hinds and Ruffett, 1971; Moss et al., 1972; Mitro and Schiebler, 1972; Hannah and Geber, 1977; Privat, 1977).

Various types of ependymal cells were examined by Brightman and Palay (1963) with electron microscopy in the region of the third and fourth ventricles, and in the area of the central canal. Tennyson and Pappas (1962) distinguish three types of cells: (1) simple ependymal cells with flattened or ovoid bases; (2) ependymal tanycytes, described by Horstmann (1954), that are characterized by long basal processes extending into the underlying neuropil, and (3) ependymal astrocytes having basal processes that branch several times. Hirano and Zimmerman (1967) described special features of the ependymal cells in the lateral ventricles of rats. Westergaard (1970) described the ependyma of the lateral ventricle in the rat, and found that ependymal cells can be divided into 3 main groups, i.e. regular cylindrical, irregular cuboidal, and irregular squamous types. In his opinion, these differences in the structure of the ependyma represent its different functions.

In her work, Millhouse (1972) presumed the ependymal cells with arch-shaped basal parts. Other ependymal cells project short or long processes that penetrate under the neighbouring ependymal cells or in the adjacent neuropil.

Generally, it is very difficult to describe an ependymal cell as the "standard type" or "normal type" because of the wide variety of ependymal cells occurring in different regions of the ventricle walls (Mitro, 1976).

As already mentioned, special types of the ependymal cells with long basal processes are termed tanycytes. They occur mostly in the central part of the third ventricle, and their processes connect the third ventricle with the external CSF space, capillaries of the median eminence and the capillaries in the basal hypothalamus. This topographic situation suggests a transport function of tanycytes between the ventricle and the brain vessels (Horstmann, 1954; Löfgren, 1959a; Millhouse, 1971; Brawer, 1972; Güldner and Wolff, 1973; Akmayev et al., 1973; Akmayev and Fidelina, 1974; Wagner and Pilgrim, 1974).

Above the ventricular surface of the ependymal cells are found the so-called supraependymal structures such as (1) formations that start out from the ependymal cells as, for example, the cell protrusions, protoplasmic globules, etc., (2) formations of undefined origin, for example, the supraependymal cells, and (3) formations that do not derive from the ependymal cells as, for example, the intraventricular nerve fibres (Schwanitz, 1969; Coates, 1973; Mesters and Breipohl, 1976; Ribas, 1977a, b).

In some parts of the ventricle system so-called intraventricular balls can be demonstrated by electron microscopy. They were described and termed "Kugeln" (Leonhardt, and Lindner, 1967) or "free supraependymal homogeneous structures" (Schwanitz, 1969) or also "ependymal bleb" (Coates, 1973).

These balls are believed to be large particles. Their connections with the cells are rarely preserved. They are probably structures similar to the cellular protrusions that can be demonstrated in various regions of the ventricle ependyma. These cellular protrusions may be secretory equivalents of the ependymal cell despite the fact that we could not find any close contact with the Golgi apparatus. It is supposed that these balls may influence the composition of the CSF. As it has been stated by Milhorat et al. (1971), the CSF is produced not only by the choroid plexus but also by the ependymal lining of the walls of the brain ventricles. The question of ependymosecretion was dealt with by Knowles (1969), Wittkowski (1968) and Booz (1975).

On the apical surface of the ependymal cells there are microvilli varying in number and distributed unevenly. There are regions where microvilli are densely packed and others where they occur only occasionally. Aggregation of microvilli in groups has been described in the ventricle system of rats (Millhouse, 1972).

Microvilli and cilia can be observed in uneven distribution on the apical surface of the ependymal cells (Brightman, 1961; Worthington and Catchcart, 1963; Dalen et al., 1971; Nakayama and Kohno, 1974). The tanycytes do not have cilia. As usual, the cilia contain 9 + 2 filaments (Brightman and Palay, 1963). Some of the cilia exhibit cytoplasmic thickenings which can be observed by scanning electron microscopy (Yamadori and Yagihashi, 1975). The length of the cilia varies, and they generally appear in small tafts (Kozlowski et al., 1973). In some ependymal regions of the ventricle system of rats, such as the lower part of the third ventricle (Schachenmayr, 1967) or the area postrema (Špaček and Pařizek, 1969), ciliated cells proved to be absent.

Between the ependymal cells neuronal elements, like nerve fibres and nerve endings, have been described in many regions of the ventricle system. The nature and fine structure of the intraventricular nerve fibres have been studied in various regions of the ventricle system in a number of species (Agduhr, 1932; Pesonen, 1940; Dierickx, 1962; Ito, 1965; Leonhardt, 1969; Arnold, 1970; Noack and Wolff, 1970; Lorez and Richards, 1973; Aghajanian and Gallager, 1975; Lorez et al., 1975; Terneaux et al., 1977). In the region of the fourth ventricle of rabbit the intraventricular nerve fibres were investigated by Leonhardt and Lindner (1967), Leonhardt and Backhus-Roth (1969). These fibres were found to be adjacent to the ependymal cells, forming synapse-like contacts on the surface of the ependymal cells.

The nerve fibres are characterized by the presence of the mitochondria and the neurotubules, 200–250 nm in diameter; their number depends on the thickness of the fibre. The origin of these fibres has not been elucidated.

The other neuronal elements within the ependyma are the nerve endings. They appear in the form of bulbous enlargements localized supraependymally containing one or more mitochondria, several 90–110 nm dense-core vesicles, and also a large number of 40–60 nm clear vesicles. According to histofluorescent and electron microscopic studies of the fourth ventricle, in the rat these supraependymal nerve terminals are serotoninergic (Lorez and Richards, 1975). Their origin is not clear. Intraventricular nerve endings were also described by Westergaard (1972) who examined the lateral ventricles of rat. Leonhardt and Prien (1968) described nerve endings in the fourth ventricle of rabbits.

The cytoplasm of the ependymal cells contains both free and membrane-associated ribosomes, but it is generally poor in rough endoplasmic reticulum. The Golgi apparatus varies in size. Mitochondria are mostly localized in the apical part of the cells. All these cell organelles were indicative of some degree of secretory activity, however, intraventricular secretory product could not be demonstrated as yet (Brightman, 1961; Brightman and Palay, 1963; Hirano and Zimmerman, 1967; Leonhardt, 1969; Mitro, 1969; Nakai, 1971; Millhouse, 1972; Leonhardt, 1974; Yamadori and Yagihashi, 1975).

The shape of the nuclei varies according to the shape of the ependymal cells. They are mostly oval, vertically oriented to the surface and contain eccentric nucleoli. In the flattened ependymal cells the elongated nuclei lie parallel to the cell surface. On the inner membrane of these nuclei there were clusters of chromatin in contrast to other types of nuclei which were surrounded only by a small amount of chromatin (Desaga, 1971).

The lateral walls of the ependymal cells seemed to be uneven owing to membrane interdigitations. The so-called zonulae occludentes and zonulae adherentes can be distinguished from the apical part of the cells. They might play a certain role in the passage of substances from the CSF (Léránth and Schiebler, 1974).

Below the ependyma the second layer of the ventricle walls can be localized. This is termed subependyma or hypependyma. When present, it separates the ependyma from the underlying periventricular nervous structures (Klinkerfuss, 1964; Blakemore, 1969; Blakemore and Jolly, 1972; Privat and Leblond, 1972; Nakai et al., 1977; Privat, 1977).

The subependyma is composed of processes and cell bodies of the astrocytes or other glial cells. In the subependyma, numerous nerve fibres are in parallel arrangement with the ventricular surface of the ependymal cells. Furthermore, at some places they form two or more systems of the astrocytic processes which cross each other or form an irregular network (Marker, 1970; Fleischauer, 1972).

In some species the basement membrane labyrinths are formed between the ependyma and subependyma (Leonhardt, 1972). Leonhardt and Desaga (1975) also described labyrinths between the ependymal cells. These basement membranes form further labyrinths with the perivascular space of the subependymal vessels. Moreover, these authors suggest transport possibilities between the subependymal basement membrane labyrinths and the pericapillary space (Desaga, 1972; Booz et. al., 1972, 1974).

Under the subependymal layer periventricular nervous tissue (gray matter) is found. The periventricular tissue has a characteristic morphological structure in various parts of the brain.

The ependymal cells that cover the circumventricular organs (CVO) have a specific form and role. It is well known that the ependymal ce llst ogether with the cells of the hypependyma form circumventricular organs. Vigh (1971) divided these organs into three groups: ependymal, hypependymal, choroidal types of the circumventricular organs. The CVO's structure has been studied by Palkovits (1965), Akert et al. (1967), Weindl et al. (1967, 1968), Phillips et al. (1974).

The tela choroidea is localized on the roof of the ventricles. The choroid plexus is formed together with the vascular folds of the pia mater. The vascular folds contain blood capillaries embedded in connective tissue (stroma). The outer surface of the choroid plexus is covered by epithelium (lamina epithelialis) derived from the ependymal lining of the ventricles. The choroid epithelium is also termed ependyma of the choroid plexus. The pial and epithelial components together are termed tela choroidea (Cancilla et al., 1966; Matano and Ishii, 1973; Peters, 1974).

In some regions the ependymal lining is penetrated by the processes of certain nerve cells. These cells do not respect the ependymal barrier that separates the nervous elements from the CSF. Their processes (endings) have special forms, and their common feature is the close connection with the CSF. Such neurons are called CSF-contacting neurons (Vigh and Vigh-Teichmann, 1973; Vigh-Teichmann and Vigh, 1974). The perikarya of these nerve cells are similar to the perikarya of other nerve cells localized in the periventricular gray matter.

A great number of histochemical studies can be found in the literature demonstrating the enzyme activities of the ependymal cells (Bartoniček and Lojda, 1964, 1966; Shuttleworth and Allen, 1966; Goslar and Bock, 1971; Luppa and Feustel, 1971; Kroon and Goossens, 1974; Sarnat et al., 1975; Firth, 1976).

Enzymes like non-specific esterase, glucose-6-phosphate dehydrogenase, nicotinamide nucleotide and nicotinamide nucleotide phosphate tetrazolium reductase generally yield a positive reaction in certain parts of the ependyma. The histoenzymological studies are more informative about the activity of the individual ependymal cells rather than about the topographic distribution of the various ependymal areas formed by the same type of ependymal cells. This means that positive reactions can be observed in the cells independently of these localizations.

Therefore, it has to be emphasized that studies of the ependyma should always be completed with the morphological characterization of the ependymal cells (Mitro, 1975).

LATERAL VENTRICLE

Topographic coordinates

In the schematic drawings of the lateral ventricle (Fig. 5) separated parts of the lateral ventricle and coronal transections of the lateral ventricles are depicted. The coronal transections are identical with those visible in the frontal sections of the brain in Fig. 2A (sections 1–32).

The lateral ventricles constitute the major portion of the ventricle system. The total ependymal surface of the lateral ventricle in a 200 g rat is about 30 mm^2 which amounts to nearly 40% of the total ventricular surface (Table 1).

The lateral ventricle can be divided into three major parts (Fig. 2): the anterior horn (sections 1–14, coordinates A3.0 mm to P0.9 mm); middle portion of the central part (sections 15–22, coordinates P1.2 mm to P3.3 mm); and posterior portion of the central part with the inferior horn (sections 23–32, coordinates P3.6 mm to P6.3 mm).

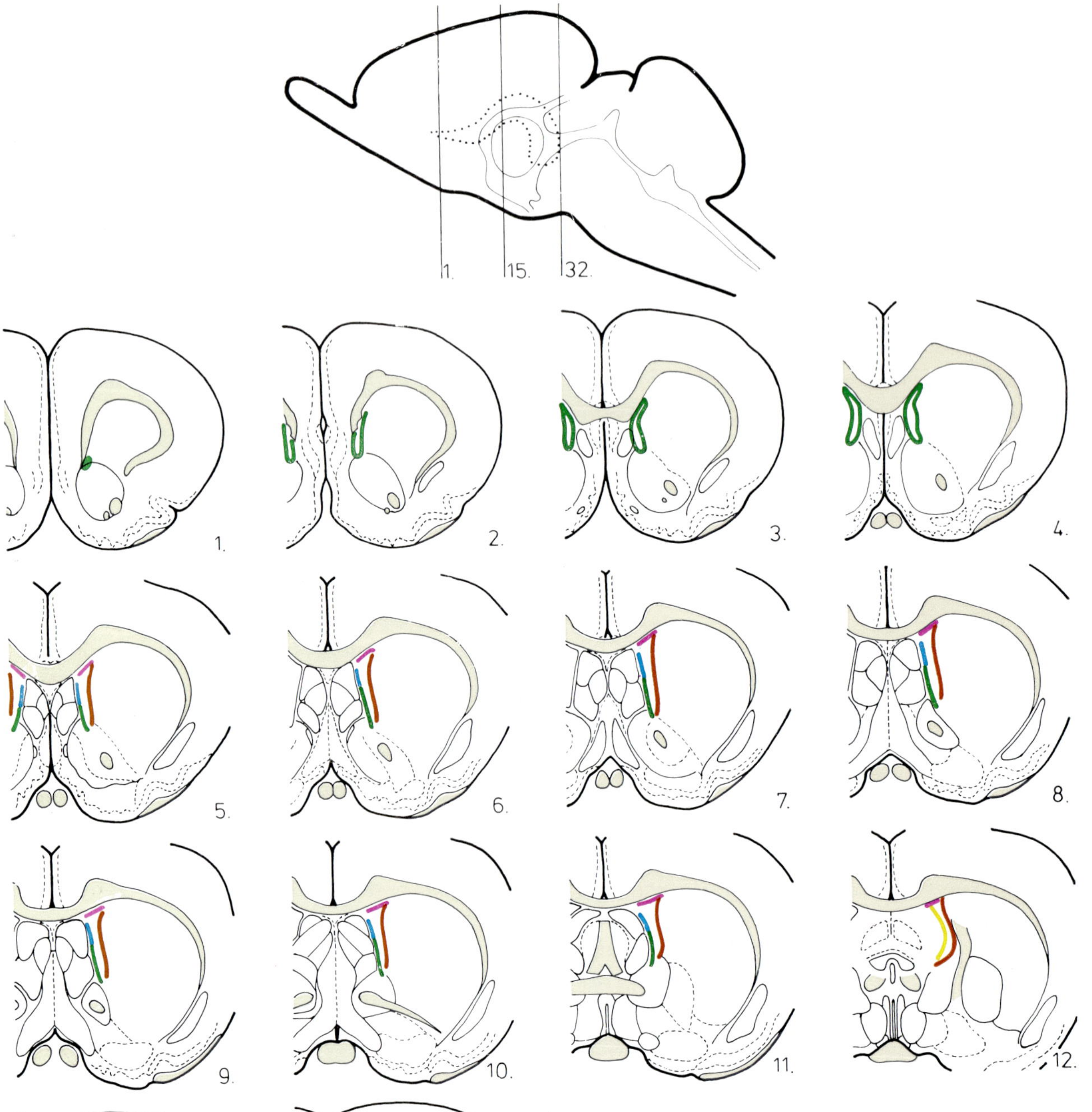

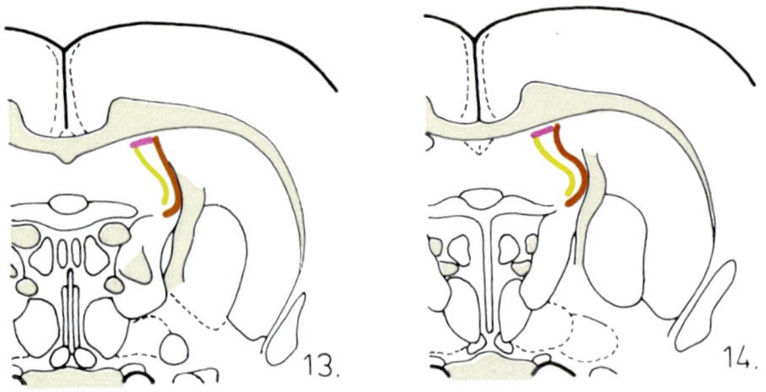

Fig. 5. Schematic drawings of coronal transections of the lateral ventricle. The different ependymal areas are indicated in different colours. The separated parts of the lateral ventricle are shown in the sagittal drawings. Green — anterior area; violet — top ependymal area; blue — dorsal part of the internal wall; yellow — transitional area of the anterior horn; brown — the external area of the anterior horn and the

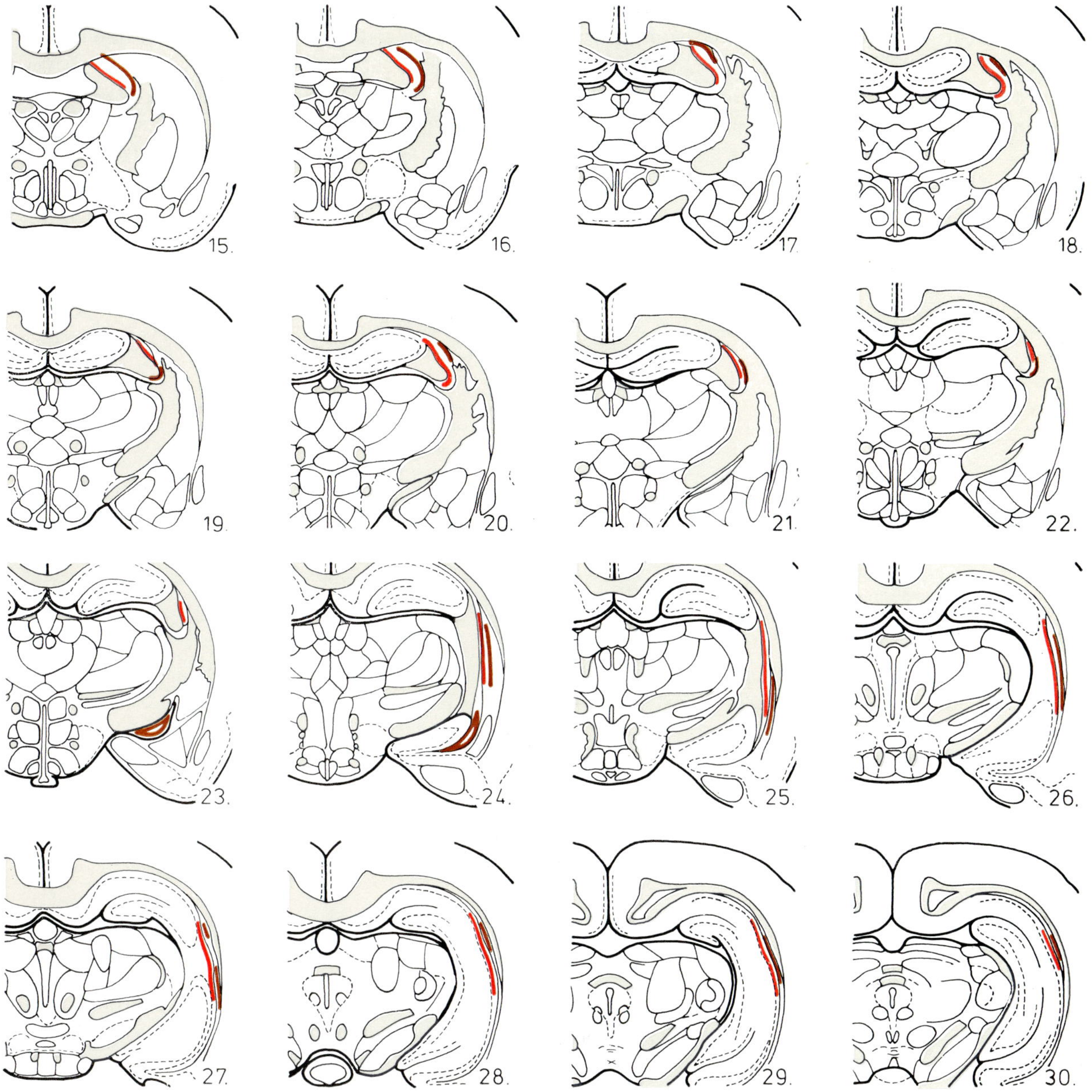

central part; red — internal area of the central part; black — inferior horn. Anterior horn (sections 1–14 = from A3.0 mm to P0.9 mm); middle portion of the central part (sections 15–22 = from P1.2 mm to P3.3 mm); posterior portion of the central part and inferior horn (sections 23–32 = from P3.6 mm to P6.3 mm)

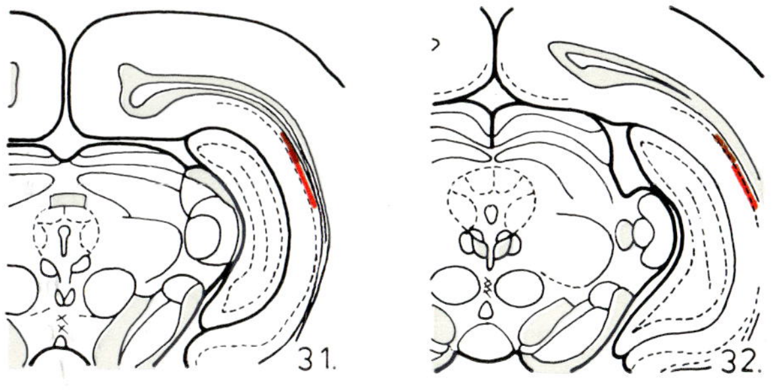

According to their ependymal structures, the regions of the lateral ventricle can be further subdivided.

The anterior horn constitutes the largest subdivision of the lateral ventricle. Its total ependymal surface is 12.6 mm², which is over 40% of the total surface of the lateral ventricle (Table 2).

The caudal portion of the central part of the lateral ventricle is much more developed than the middle one. Its ependymal surface is nearly 12 mm² consisting of an internal and an external wall (Fig. 5). The surface of the internal wall is larger than that of the external wall (Table 2).

The inferior horn is rather small in the rat. Its ependymal surface is not more than 1 mm² (Table 2).

Table 2

LATERAL VENTRICLE

	Extension*		Surface mm²	Surface fraction % 100% = *30.89 mm²*
	Rostral	Caudal		
	mm			
Anterior horn	A3.0	P0.9	*12.62*	*40.9*
anterior area	A3.0	0	2.95	9.6
top ependymal area	A1.8	P0.9	1.64	5.3
upper internal wall	A1.8	0	1.07	3.5
external wall	A1.8	P0.9	6.08	19.7
transitional area	P0.3	P0.9	0.88	2.8
Middle portion of the central part	P1.2	P3.3	*5.34*	*17.3*
internal wall	P1.2	P3.3	2.48	8.0
external wall	P1.2	P3.3	2.86	9.3
Caudal portion of the central part	P3.6	P6.0	*11.95*	*38.7*
internal wall	P3.6	P6.0	6.73	21.8
external wall	P3.6	P6.3	5.22	16.9
Inferior horn	P3.6	P3.9	*0.98*	*3.2*

* 0 = bregma level, A = rostral, P = caudal to the bregma.

Within the anterior horn an anterior area, a top ependymal area, upper internal and external walls and a small transitional area can be distinguished (Fig. 5). The external wall forms nearly the half of the wall surface of the anterior horn, which, according to surface territory, is followed by the anterior area. The surface of the other regions is less than 2 mm² (Table 2).

The middle portion of the central part of the lateral ventricle is smaller, less than one-fifth of the total surface of the lateral ventricle. Its ependymal wall can be divided into an external and an internal part. They are approximately of the same size (Table 2).

The caudal portion of the central part of the lateral ventricle is much more developed than

Structure of the ependymal areas of the lateral ventricle

The initial part of the lateral ventricle, called the anterior area, begins as an aggregation of nerve and glial cells. Backwards, the ventricle acquires the form of a slot. The ependyma is monolayered but in some small areas of the ventricular lining it becomes multilayered. The ependymal cells are cuboidal with round nuclei (Fig. 6A). In some places subependymal cells occur with elongated nuclei (Fig. 6B).

After the appearance in frontal sections of the trunk of the corpus callosum the ventricle acquires a polygonal shape. The ependyma covering the corpus callosum (top ependymal area) is monolayered. The ependymal cells are flat or

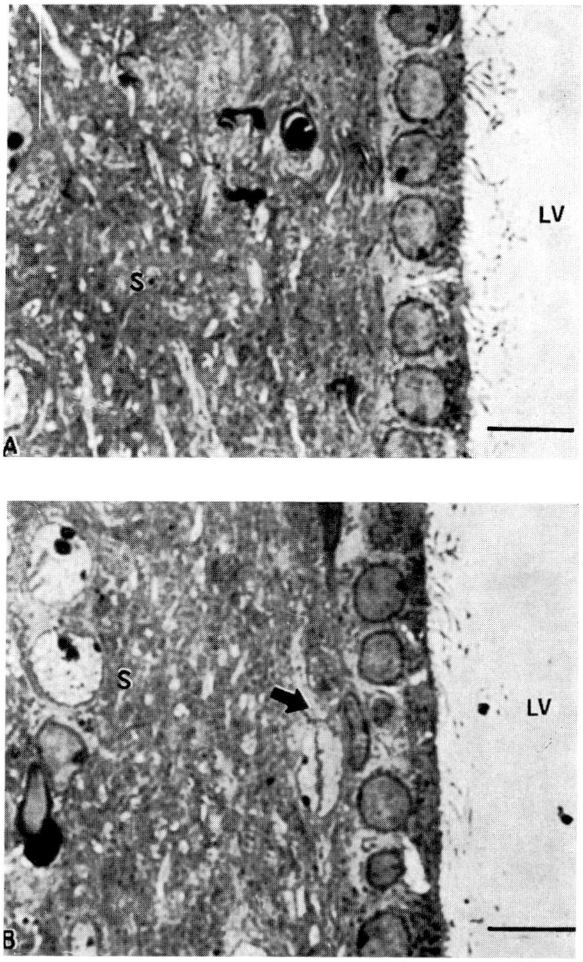

Fig. 6. Cuboidal ependyma from the anterior area of the anterior horn. **A** — regular single layer of cuboidal cells; **B** — cuboidal ependyma with other cell types. Subependymal cells with elongated nuclei are indicated by arrow. LV — lateral ventricle, S — septal cells and neuropil. Bar scale = 10 μm

cuboidal with round to oval nuclei. These are localized at irregular distances from each other (Fig. 7).

The upper internal wall of the anterior horn is formed by densely packed cuboidal, at some sites, by columnar ependymal cells with round nuclei. This part of the lateral ventricle has a close topographic connection with the septum pellucidum. The ependymal cells are ciliated and are placed directly on the neuropil or in small groups of subependymal cells (Fig. 8). The subependymal cells have oval to longitudinal nuclei in parallel localization with the luminal surface of the ependyma.

The external wall of the lateral ventricle which is connected with the caudate nucleus is

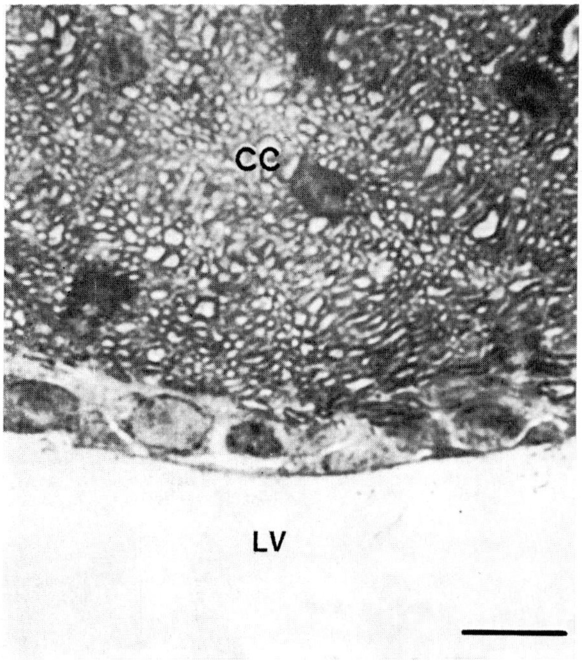

Fig. 7. Cuboidal and flat ependymal cells from the top ependymal area of the anterior horn of the lateral ventricle (LV). Under the ependymal layer cross section of the fibres of the corpus callosum (CC) are seen. Bar scale = 10 μm

made up of flat or cuboidal ependymal cells with oval nuclei lying upon the subependymal cells (Fig. 9A, B, C). The thickness of subependymal cells is uneven. The external wall in the lower part of the anterior horn is covered by

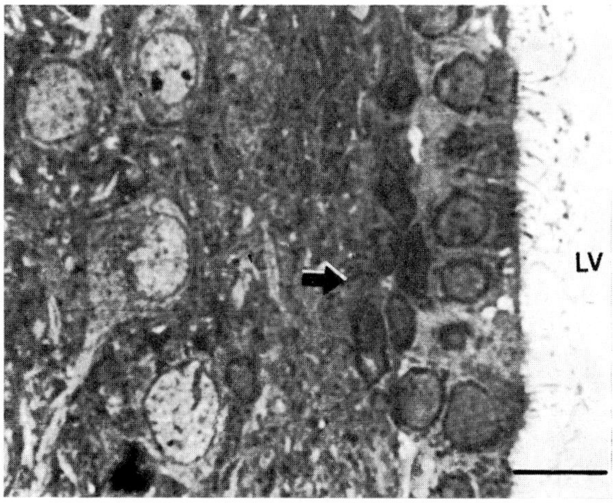

Fig. 8. Cuboidal and columnar ependymal cells from the upper area of the internal wall of the anterior horn (LV). Blue area in Fig. 5. The subependymal cells with small elongated nuclei can be easily distinguished from the large cells of the dorsal septal nucleus. Bar scale = 10 μm

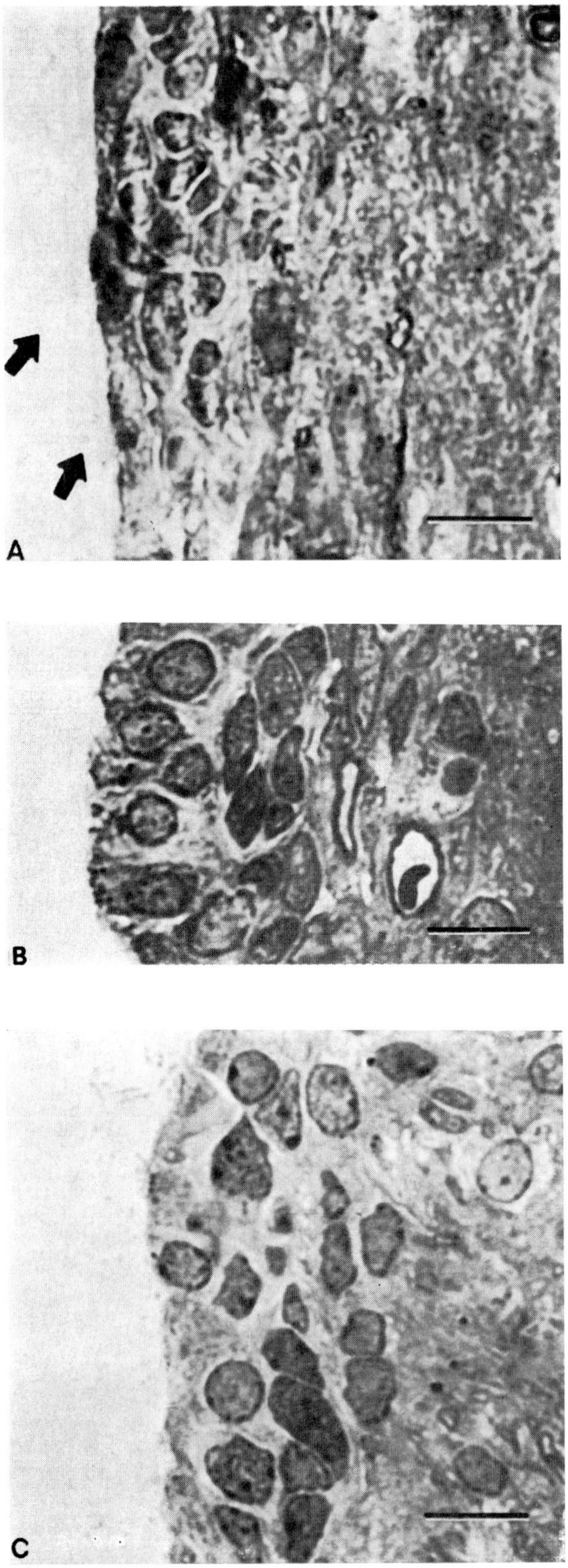

Fig. 9. Variations of ependymal cells in the external area of the anterior horn (brown area in Fig. 5). **A** — flat cells with cilia (→), well-developed subependymal cells; **B** — cuboidal ependyma; **C** — irregular ependyma. Bar scales = 10 μm

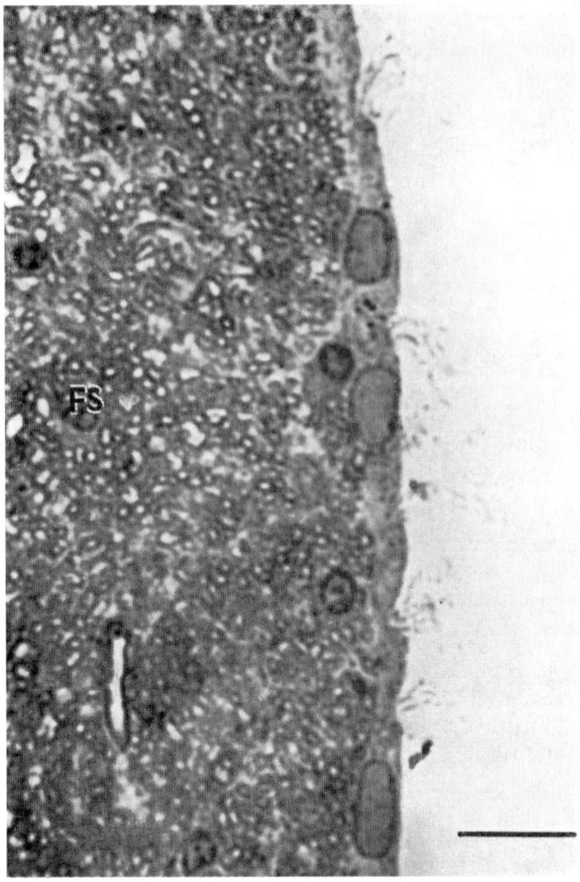

Fig. 10. Extremely flat ependymal cells from the transitional area of the anterior horn of the lateral ventricle (yellow in Fig. 5). There are no subependymal cells between the ependyma and the septal fimbria (FS). Bar scale = 10 μm

multilayered ependyma. The nuclei of the cells are round or oval. The cell surface is ciliated.

In the region of the interventricular foramen (Fig. 5, sections 12, 13, 14) the inner wall is formed by extremely flat ependymal cells (Fig. 10). This region is called the transitional area of the anterior horn.

As to the interventricular foramen, the connection of the lateral ventricle with the third one is in fact a narrow canal so that the term "canalis interventricularis" would be appropriate as well. The wall of this canal is lined with monolayered cuboidal ependyma which is formed also at both ends of the "canals".

Backwards, the lumen of the lateral ventricle continues as the middle portion of the central part. The medial and lateral walls can be distinguished not only by their topography but also by their ependymal structures. The internal (medial) wall is formed of flat, irregular ependyma. The shape of the cell nuclei, located at uneven distances from each other, is mostly oval (Fig. 11A, B).

The external (lateral) wall is formed in some places by cuboidal, in others by flat, irregular ependyma. The cell nuclei are round to oval (Fig. 12A). Some ependymal cells, with their lateral processes, cover the neighbouring ependymal cells (Fig. 12B). Under the ependyma subependymal cells are localized irregularly. They possess dark, longitudinal nuclei (Fig. 12C).

The caudalmost part of the lateral ventricle is the posterior portion of the central part. The medial and lateral walls of the posterior portion of the central part are covered by ependyma that looks like the ependyma in the middle portion of the central part.

The walls of the inferior horn are lined with monolayered flat to cuboidal ependyma, the ependymal cells have cilia and round to oval nuclei (Fig. 13A, B). The lower part of the inferior horn has larger ependymal cells, assembled in groups of cells (Fig. 13C).

Above the posterior part of the lateral ventricle fusions of the ventricle walls may occur. Westergaard (1970) termed these fusions coarctationes ventriculi.

Backwards, the lumen of the ventricle becomes narrower and the ependymal lining gradually thinner. The ependymal cells are flat to cuboidal with round or oval nuclei. The distances between the nuclei are uneven. The cell nuclei are somewhat smaller in the internal as compared to those in the external wall.

The lateral ventricle ends in the form of a flock of ependymal cells with small round nuclei.

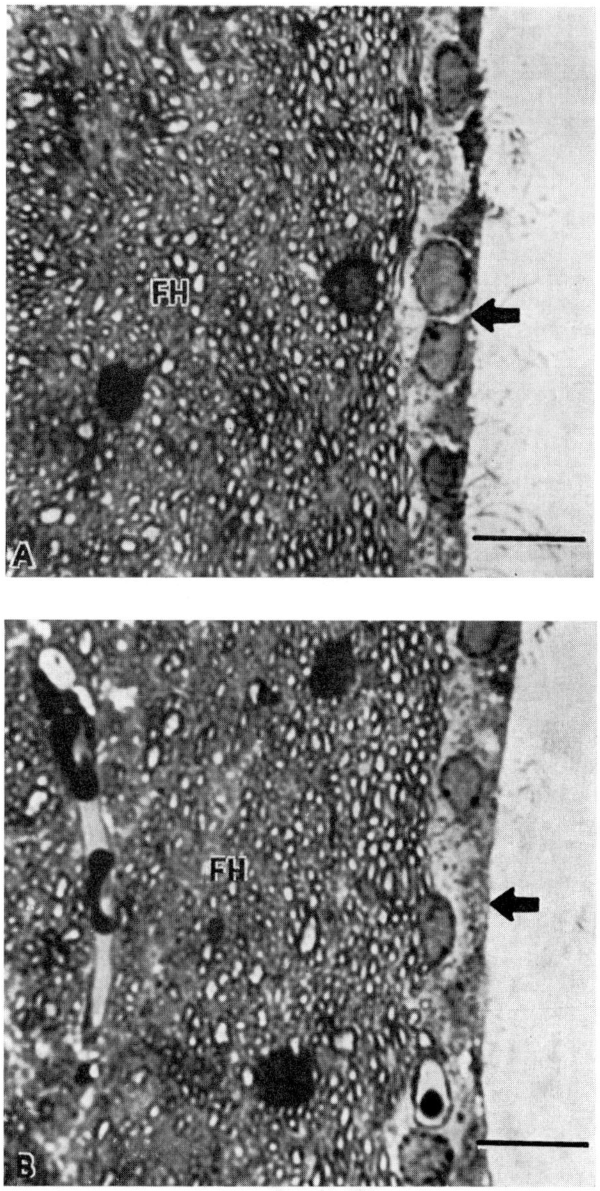

Fig. 11. **A** — flat, irregular ependyma from the medial wall of the central part of the lateral ventricle; **B** — irregular ependymal cells with basally located nuclei (→). This is a red area in Fig. 5. FH — cross sections of the fibres of the hippocampal fimbria. Bar scales = 10 μm

Cell groups in the vicinity of the lateral ventricle

The anterior horn is bordered medially by the septum, and laterally by the caudate nucleus. This part of the ventricle system has a great importance in experimental brain research since the intraventricular injections are applied generally to the caudal part of the anterior horn of the lateral ventricle.

A close topographic correlation seems to exist between the different ependymal areas and related periventricular structures. The lateral septal nucleus is covered by the cuboidal ependyma of the anterior area (Fig. 2B/3–9). The dorsal septal nucleus is connected with another type of ependyma; its ventrical surface is similar to the upper internal wall (Fig. 2B/3–9).

The small upper ependymal area lies immediately under the ventral surface of the corpus callosum. This is the only place where the lateral

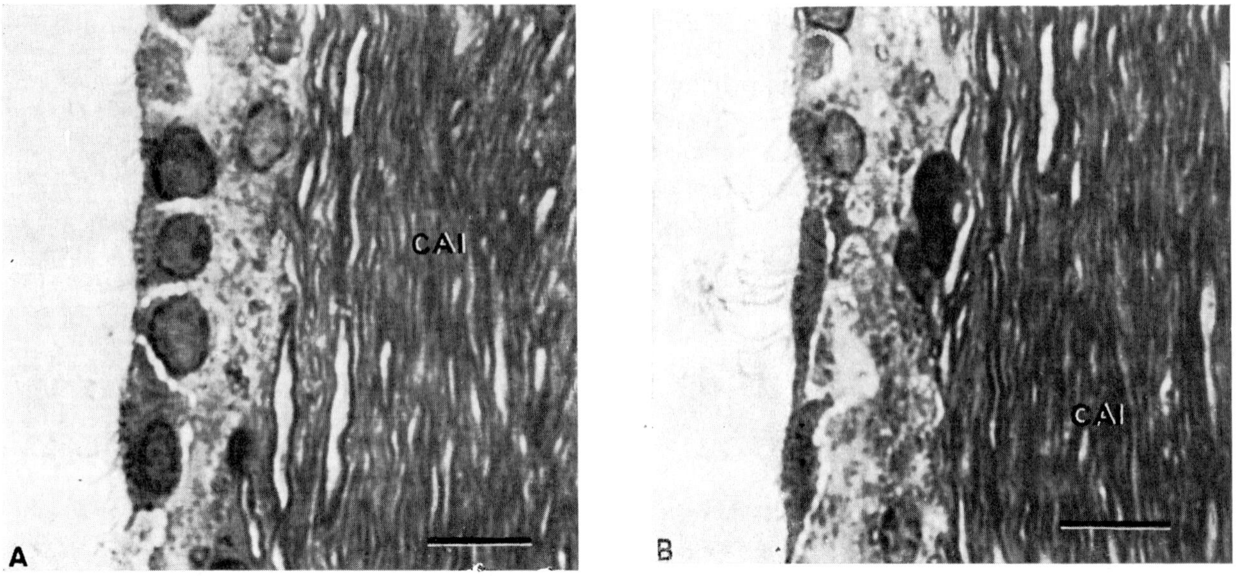

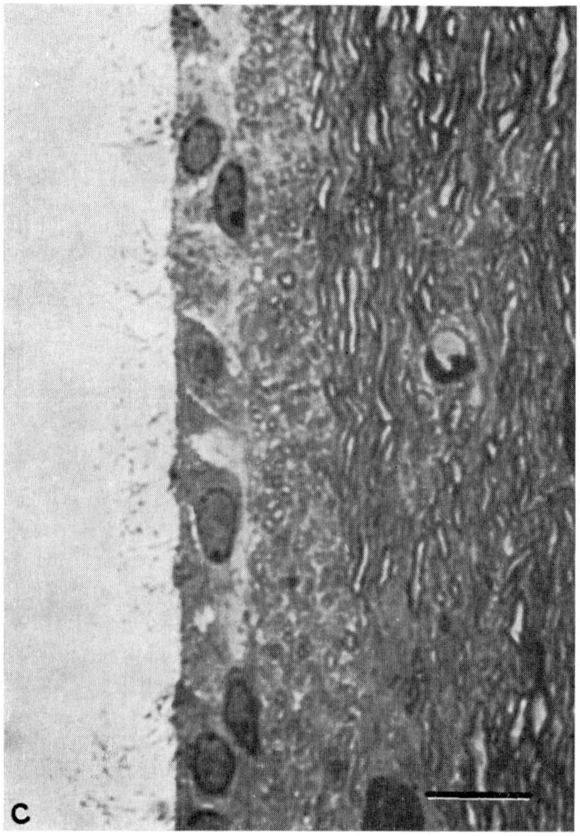

Fig. 12. **A**, **B**, and **C** — irregular (flat to cuboidal) ependyma from the lateral wall of the central part of the lateral ventricle (brown in Fig. 5). Subependyma, varying in thickness, can be seen between the ependyma and the longitudinal sections of fibres running in the internal capsule (CAI). Bar scales = 10 μm

ventricle is devoid of subependymal periventricular cellular structures (Fig. 2B/3–12).

The external wall of the anterior horn is uniform regarding both the ependymal and periventricular structures. A relatively large portion of the ventricular wall belongs to the caudate nucleus (Fig. 2B/3–12).

The septal fimbria and the hippocampal fimbria caudalwards, are covered by the special ependyma (transitional area) of the internal wall of the anterior horn (Fig.2B/10–12). Other brain areas, like the nucleus accumbens, stria terminalis, nucleus of the stria terminalis, are just touched by the anterior horn without any significant topographic connection to the ependyma.

The middle portion of the central part of the lateral ventricle can be divided into an external and an internal wall. They are covered by different types of ependyma. The internal wall is found on the surface of the hippocampal fimbria, while the external ependyma is connected to both the cells of the caudate nucleus and the fibres of the internal capsule (Fig. 2B/13–20). The ventral tip of this portion of the ventricle is also connected to the fibres of the stria terminalis.

While the ependyma of the external wall itself in the caudal portion of the ventral part of the

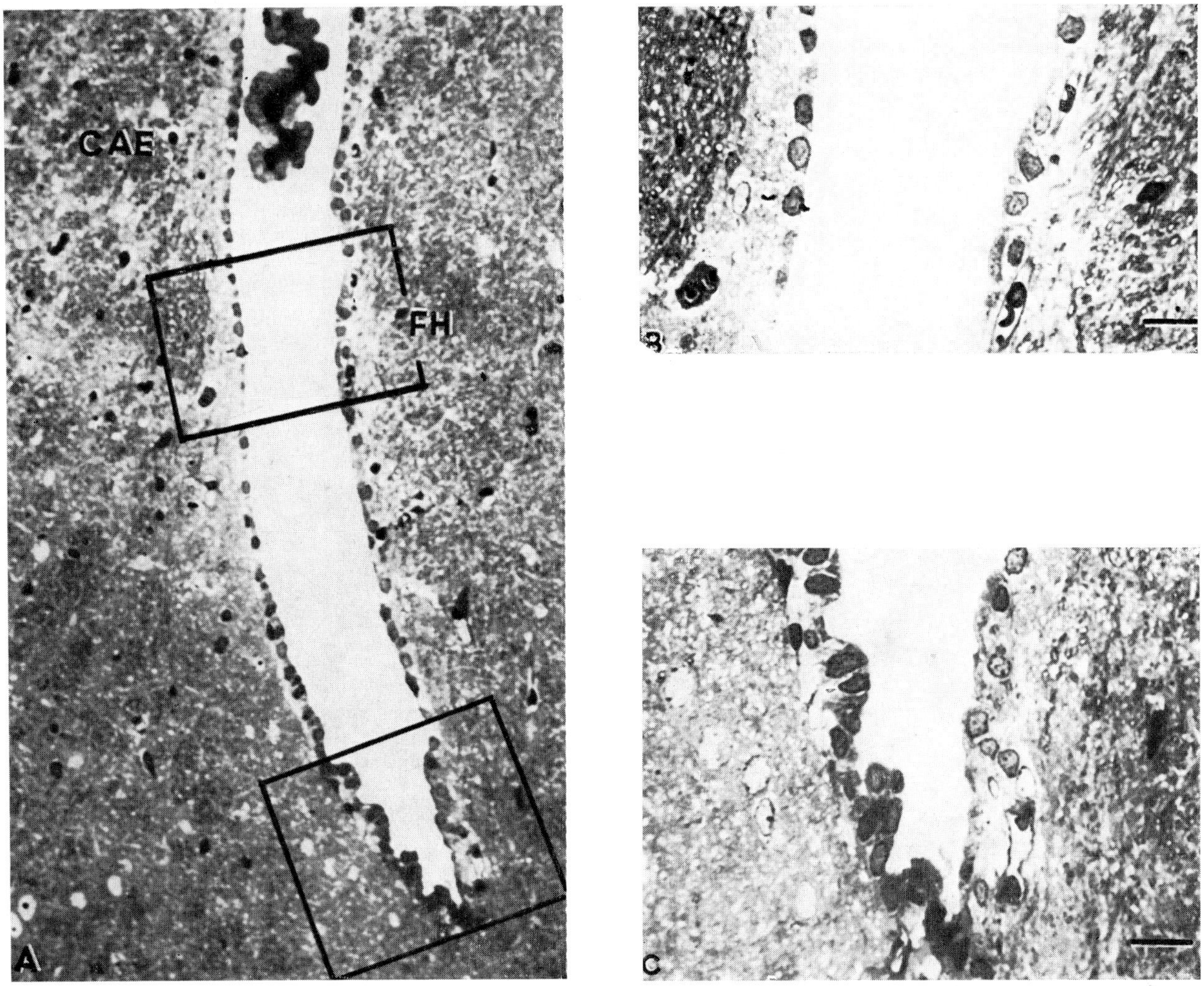

Fig. 13. Inferior horn of the lateral ventricle. **A** — toluidine blue staining. Squares represent **B** and **C**. **B** — cuboidal to flat single ependymal cell layer on the medial and lateral walls; **C** — columnar, multilayered ependyma from the ventral tip of the inferior horn. CAE — external capsule; FH — hippocampal fimbria. **A** — magnification × 100; bar scales **B** and **C** = 20 μm

lateral ventricle has a structure similar to that found in the middle portion, the subependymal structures are different. The ependyma in these regions covers the internal surface of the external capsule (Fig. 2B/21–30). The internal wall follows the fimbria hippocampi caudally as far as P4.2 mm. From this point, it can be found on the external surface of the middle portion of the hippocampus (Fig. 2B/24–29).

The inferior horn of the lateral ventricle is surrounded by the cells of the ventral hippocampus and the amygdala without any topographic or structural relation to the ependyma (Fig. 2B/21–22).

The third ventricle is the second largest of the brain ventricles. Its total ependymal surface is 21.28 mm^2 in the rat (Table 1).

It can be divided into a dorsal and a ventral part.

The dorsal part is located above the thalamus and constitutes the ventricular system of the epithalamic region. It is smaller than the ventral part. The total ependymal surface is 8.4 mm^2 (in 200 g rats), which is nearly 40% of the total ependymal surface of the third ventricle (Table 3).

The ventral part, located in the hypothalamus is larger, the ependymal surface exceeds 12 mm^2 (60.4%). It consists of many smaller areas (Table 3).

Table 3

THIRD VENTRICLE

	Extension*		Surface mm^2	Surface fraction % 100% = *21.28 mm^2*
	Rostral	Caudal		
	mm			
Dorsal part	P0.3	P6.0	**8.43**	**39.6**
Triangular recess + interventricular foramen	P0.3	P0.9	*0.34*	*1.6*
Epithalamic ependyma	P0.9	P4.2	*5.38*	*25.3*
floor			4.49	21.1
side wall			0.89	4.2
Pineal recess	P4.5	P6.0	*2.71*	*12.7*
Ventral part	A0.3	P4.5	**12.85**	**60.4**
Supraoptic recess	A0.3	P0.3	*0.94*	4.4
floor			0.08	0.4
side wall			0.86	4.0
Anterior hypothalamic portion	P0.3	P2.1	*5.91*	*27.8*
Tuberal portion	P2.1	P3.6	*4.71*	*22.1*
"recess organ"	P2.1		0.05	0.2
roof of the median eminence	P2.4	P3.6	0.57	2.7
tanycyte ependyma	P2.4	P3.6	1.84	8.6
"light cell zone"	P2.4	P3.6	1.04	4.9
cuboidal ependyma	P2.4	P3.6	1.21	5.7
"Connecting part"	P3.9	P4.2	0.58	2.7
Inframamillary recess	P3.9	P4.5	*0.71*	*3.3*
roof of the recess			0.25	1.2
tanycyte ependyma			0.46	2.2

* 0 = bregma level, A = rostral, P = caudal to the bregma.

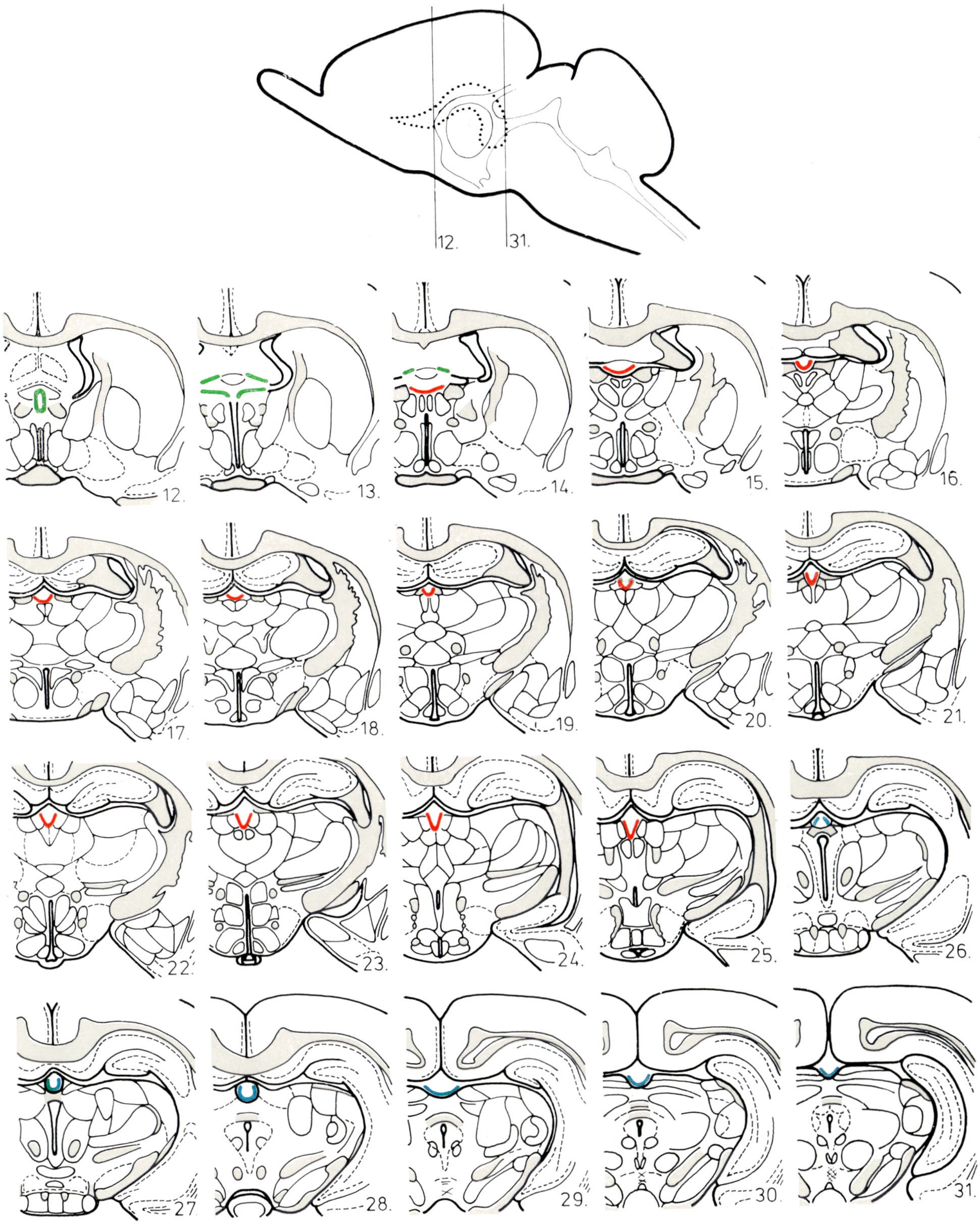

Fig. 14. Schematic drawings of the frontal sections from the ependymal areas of the dorsal part of the third ventricle. The dorsal part of the third ventricle is shown in the sagittal drawing. Cuboidal ependyma from the triangular recess and around the interventricular foramen (green, sections 12–14; P0.3 mm to P0.9 mm). Epithalamic ependyma (red, sections 14–25 = from P0.9 mm to P4.2 mm). Ependyma of the pineal recess (blue, sections 26–31; P4.5 mm to P6.0 mm)

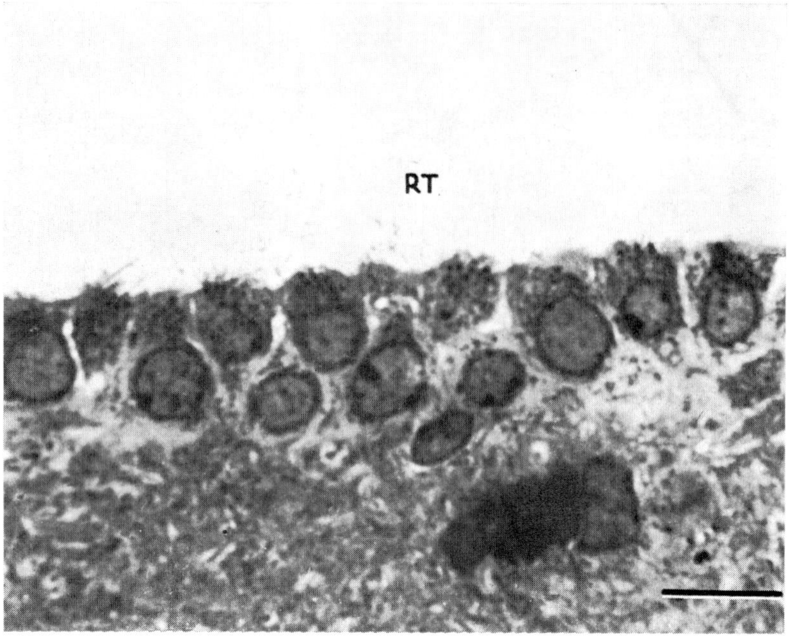

Fig. 15. Mono- to multilayered ependyma from the wall of the triangular recess (RT). Bar scale = 10 μm

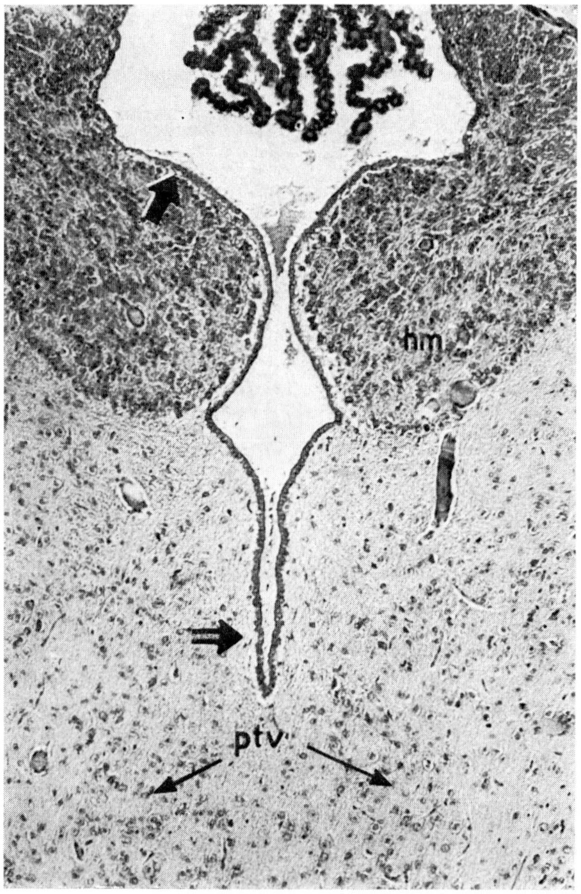

Fig. 16. Frontal section of the dorsal part of the third ventricle at the level of the medial habenular nucleus (hm). Epithalamic floor of the third ventricle (→); side wall of the epithalamic ependyma (⇒); ptv – periventricular thalamic nucleus. Magnification: ×100

Dorsal part of the third ventricle

Topographic coordinates

The dorsal part of the third ventricle extends from section 12 to section 31 (P0.3–6.0 mm) (Fig. 14). This part is localized above the massa intermedia, mainly in the epithalamic region.

We divided this part into the following ependymal areas: (1) cuboidal ependymal area of the triangular recess and around the interventricular foramen; (2) epithalamic ependyma, and (3) ependyma of the pineal recess. The different ependymal areas are demonstrated in different colours in schematic drawings of coronal sections (Fig. 14).

The largest of these areas is the epithalamic ependyma (5.38 mm^2) that can be divided into a large "floor" and a smaller side wall (Table 3). The ependymal surface of the pineal recess is smaller (2.7 mm^2) than the surface of the triangular one, while that of the interventricular foramen is very small (Table 3).

Structure of the ependymal areas of the dorsal part of the third ventricle

The beginning of the dorsal part of the third ventricle is formed by the triangular recess (Fig. 14, section 12). The recess begins as a flock of ependymal cells in the surrounding tissue of the septum pellucidum. Then it starts to have a lumen and protrudes in short folds. The recess is bordered from below by the anterior commissure. The walls are lined with cuboidal ependymal cells with round nuclei (Fig. 15). In some places the ependyma is multilayered.

The region of the interventricular foramen comprises a small segment at the mouth of the interventricular foramen (Fig. 14, section 13). The roof of this ventricle segment is formed by the subfornical organ (SFO). It is covered by flat ependymal cells having round or oval nuclei. This segment continues downwards into the ventral part of the third ventricle. The side walls are lined with monolayered cuboidal ependyma with cilia and round nuclei. On the sides of the SFO, the ventricular walls are formed of multilayered ependyma. Downwards, the ependyma becomes gradually thinner.

The habenular region occupies the epithalamus from the end of the interventricular foramen down to the beginning of the subcommissural organ (Fig. 14, sections 14–26); it is covered by the epithalamic ependyma. Various types of ependyma were found covering the floor and the side wall of the dorsal part of the third ventricle (Fig. 16). The ependyma of the floor is formed of flat, cuboidal, and cuboidal irregular ependyma which is folded in some places (Fig. 17A, B, C). The ependyma from the side wall of the epithalamic region, localized over the habenular nuclei, varies from cuboidal to flattened. The cell nuclei are arranged at irregular distances from each other. The ependymal cells have cilia. The nerve cells of the habenular nuclei are in some places separated from the ependyma by a thick layer of myelinized fibres (Fig. 18A). Occasionally, there are contacts between the ependymal and nerve cells (Fig. 18B).

Above the posterior commissure the lumen penetrates dorsally into the pineal gland; it is called the pineal recess (Fig. 14, sections 26–31). The region above the beginning of the posterior commissure contains a part of the subcommissural organ (pars supracommissuralis). More dorsally, the pineal recess is localized above the subiculum (Fig. 19A). The side walls of the pineal recess are covered in some places by regular, in other places by irregular mono- or multilayered ependyma. The ependymal cells are

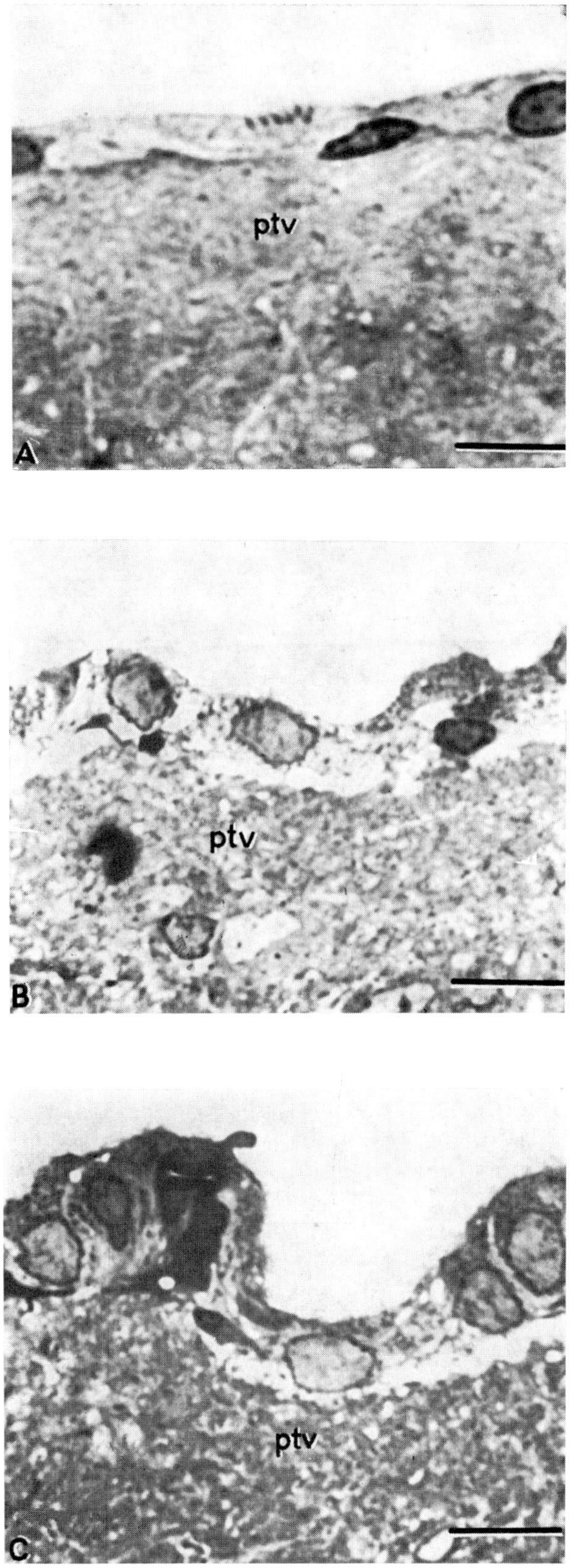

Fig. 17. Various types of ependymal cells from the epithalamic floor of the third ventricle (above the periventricular thalamic nucleus, ptv). **A** — flat; **B** — cuboidal; **C** — cuboidal ependyma mixed with other cell types. Bar scales = 10 µm

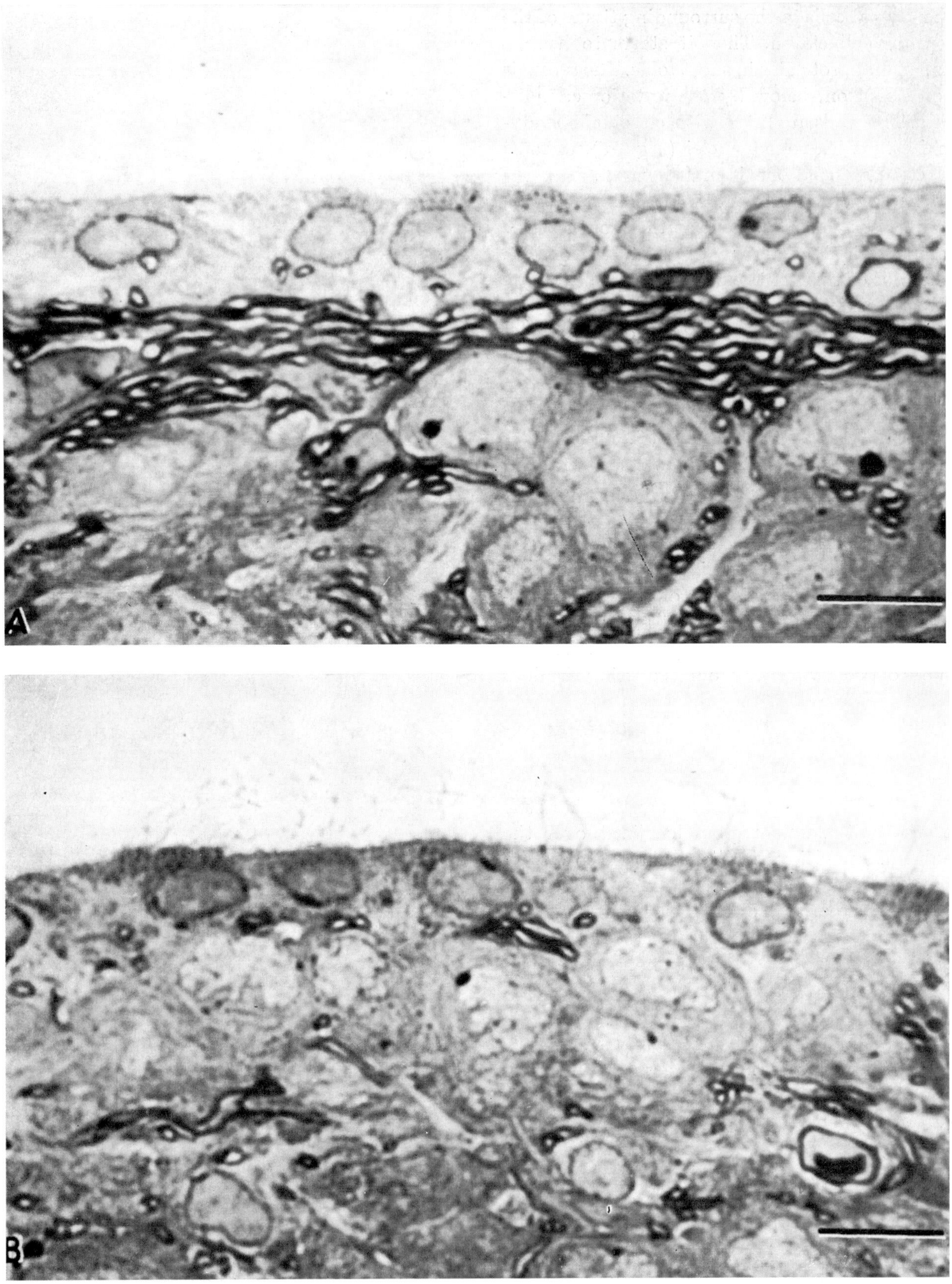

Fig. 18. Monolayered cuboidal ependymal cell from the side wall of the epithalamic region. Close topographic contact between the ependyma and myelinated fibres (**A**) and ependyma and habenular cells (**B**). Bar scales = 10 μm

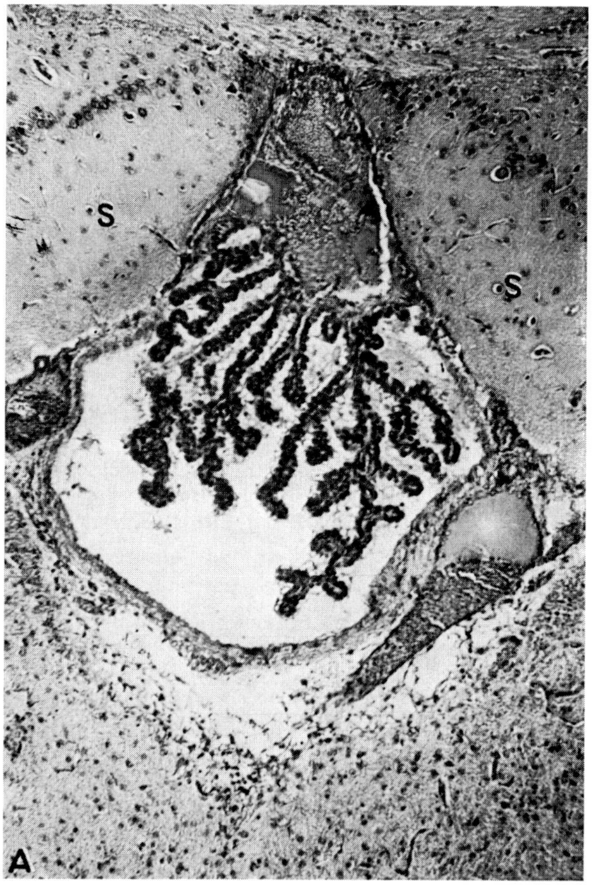

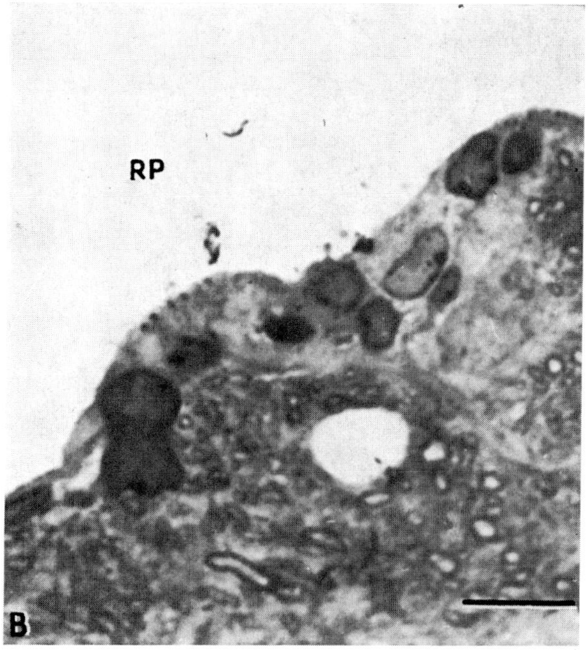

Fig. 19. **A** — frontal section of the pineal recess. Haematoxylin–eosin staining. S — subiculum. Magnification: ×100. **B** — various forms of ependymal cells cover the pineal recess (RP). Bar scale = 10 μm

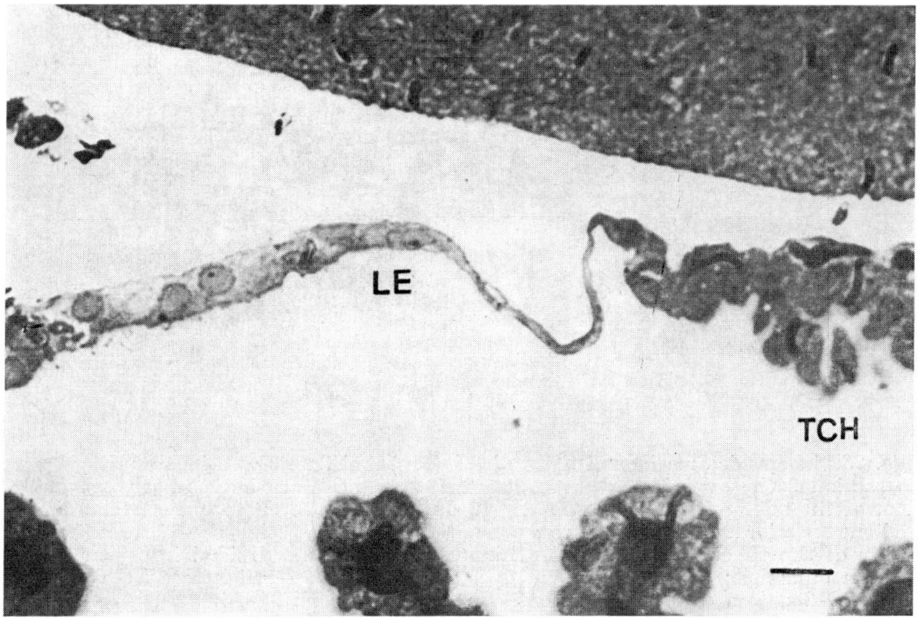

Fig. 20. Fine transitional layer between the epithelial lamina (LE) and the tela choroidea (TCH). The epithelial lamina is made up of cuboid ependymal cells. Bar scale = 10 μm

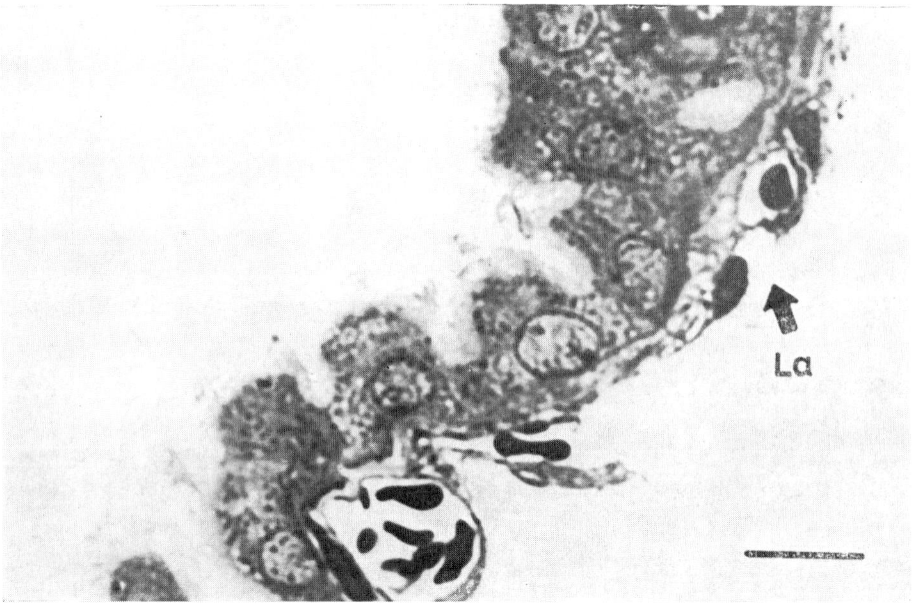

Fig. 21. Individual ependymal cells on the surface of the lamina vasculosa (La). Bar scale = 10 μm

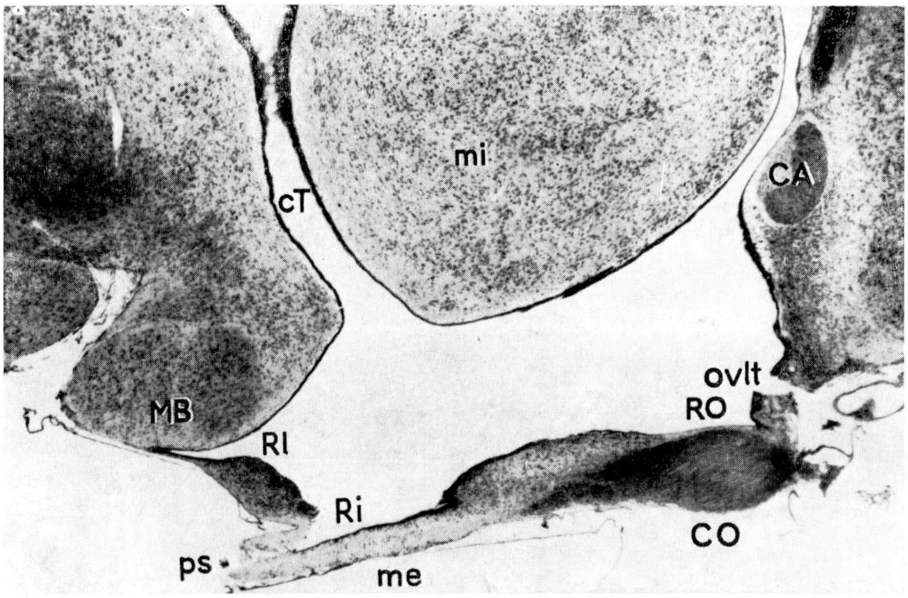

Fig. 22. Midsagittal section of the ventral part of the third ventricle. Luxol fast blue cresyl violet staining. Magnification: ×19. For abbreviations see list

⟶

Fig. 23. Schematic drawings of the sagittal (top) and frontal sections of the ventral part of the third ventricle. In section 9 the ovlt, in section 18 the "recess organ" are indicated by black. Orange — flat ependyma on the floor of the supraoptic recess (sections 10–11, from A0.3 mm to 0 mm). The side walls of the supraoptic recess and the anterior hypothalamus are represented by green (sections 10–18 from A0.3 mm to P2.1 mm). The ependymal lining of the tuberal portion of the third ventricle can be divided into four regions; violet — ependyma of the median eminence; blue — tanycyte ependyma; yellow — "light cell" zone; green — cuboidal ependyma (sections 19–23, from P2.4 mm to P3.6 mm). In sections 24–25 (P3.9 mm to P4.2 mm) the ependyma of the transitional zone is indicated by brown. The "tanycyte ependyma" can be followed on the side walls of the inframamillary recess indicated in blue in sections 24–26 (P3.9 mm to P4.5 mm). At the beginning on the roof of inframamillary recess (section 24, P3.9) "light cells" are visible (yellow). In sections 25 and 26 (4.2 mm to P4.5 mm) the roof of the inframamillary recess is made up of cuboidal ependyma (green)

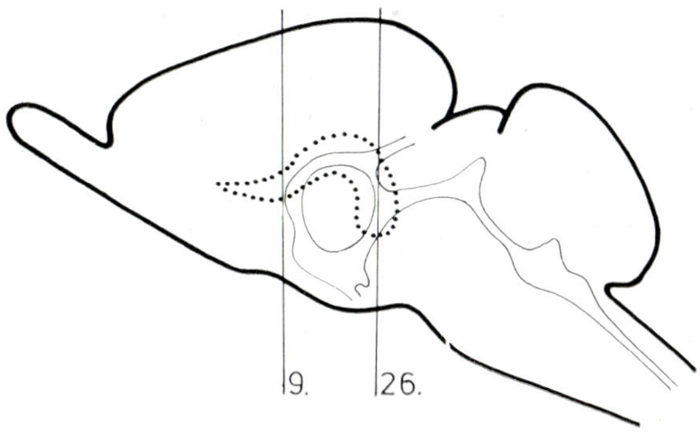

9.
26.

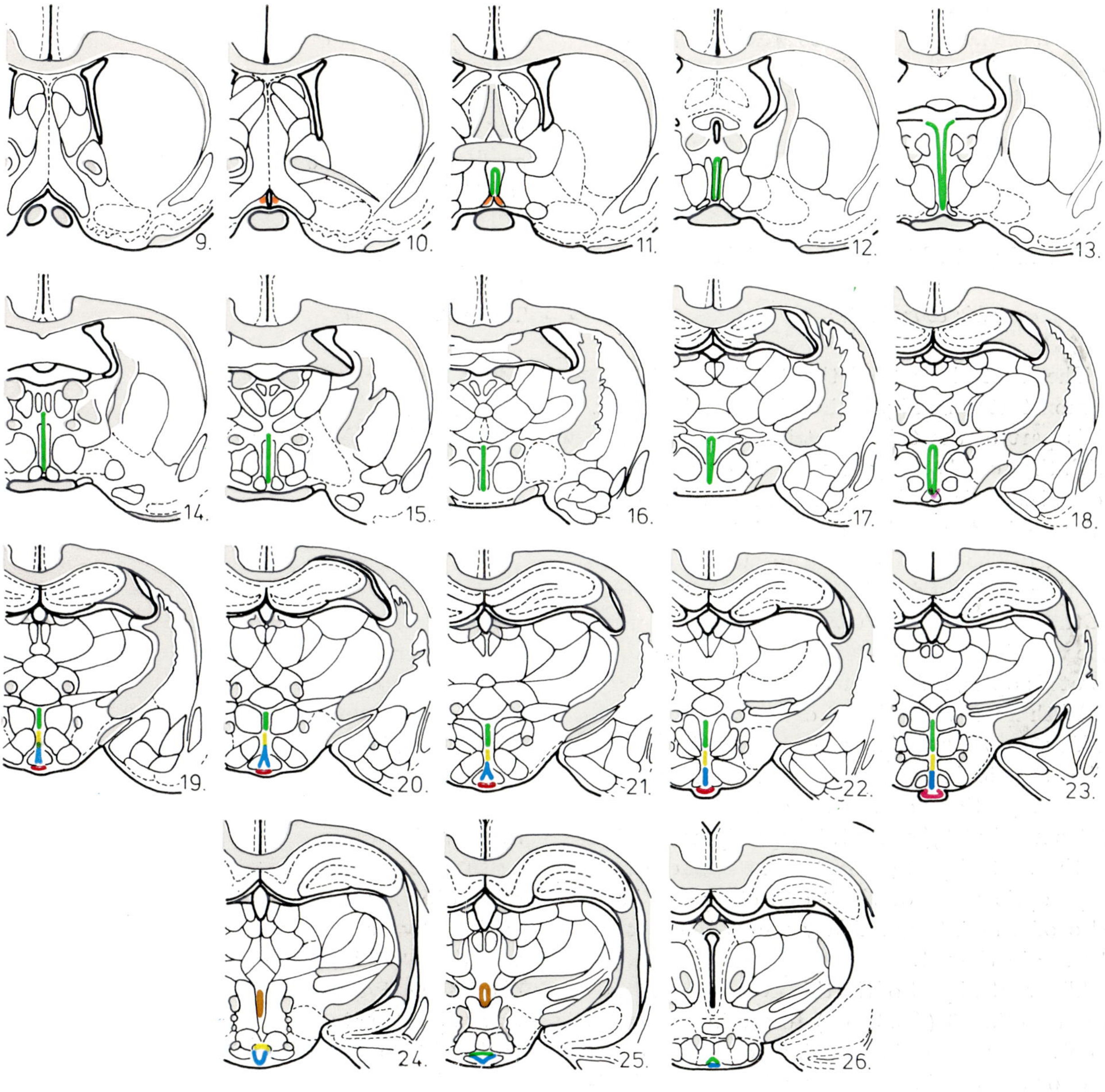

9.
10.
11.
12.
13.
14.
15.
16.
17.
18.
19.
20.
21.
22.
23.
24.
25.
26.

cuboidal or flattened. They form groups intruding into the surrounding neuropil. The cell nuclei are of unequal size, ovally shaped, and oriented either perpendicularly or parallel with the cell surface (Fig. 19B).

In the upper part of the epithalamic region, the ependyma extends as the epithelial lamina on the vascular folds of the pia mater (Fig. 20). The vascular folds are named choroid plexus. The epithelial and vascular laminae result in the formation of the tela choroidea. Ependymal cells on the lamina vascularis are situated in one row and join only at their basal parts (Fig. 21). These cells represent a special type of ependyma.

Ventral part of the third ventricle

Topographic coordinates

The ventral part of the third ventricle is demonstrated in Fig. 22 in midsagittal section. This part of the third ventricle communicates with the dorsal one by means of two canals. The rostral canal is localized above the anterior commissure in the region of the interventricular foramen. The caudal canal is formed by the connecting part of the third ventricle in the region of the cerebral aqueduct. At the ventral base of the ventricle lumen, three recesses can be demonstrated: optic, infundibular and inframamillary.

Within the ventral part of the third ventricle different ependymal areas can be distinguished. The ventricular wall inside the anterior and tuberal portions of the hypothalamus constitutes nearly 83 % of the total ependymal surface of the ventral part (Table 3). The surface of the supraoptic recess of the "connecting part" and also of the inframamillary recess is less than 1 mm^2 in the rat.

The ventral part of the third ventricle extends from section 9 (A0.6 mm) to section 26 (P4.5 mm). In schematic drawings of the frontal sections the various ependymal areas are indicated in different colours (Fig. 23).

The most rostral portion of the third ventricle is the supraoptic recess. Flat ependyma covers the floor (sections 10–11, from A0.3 mm to 0 mm). This area is very small (0.08 mm^2). The cuboidal ependyma on the side walls of the supraoptic recess continues into the same ependyma of the anterior hypothalamus (sections 10–18 from A0.3 mm to P2.1 mm; Fig. 23). This type of the ependyma covers about the half of the ventral part of the third ventricle.

The ependyma of the tuberal portion of the third ventricle can be found in sections 18–23, from P2.4 mm to P3.6 mm (Fig. 23). This area is nearly 5 mm^2 (Table 3) and can be subdivided into many other typical areas (see Fig. 36).

The ependyma of the transitional zone can be seen in section 24 (P3.9 mm) and section 25 (P4.2 mm). The inframamillary recess is small (0.71 mm^2) but rather elongated. It can be demonstrated in sections 24 to 26 (from P3.9 mm to P4.5 mm).

The rostrocaudal extensions and the size of the ependymal surfaces of the third ventricle (both dorsal and ventral parts) are indicated in Table 3.

Structure of the ependymal areas of the ventral part of the third ventricle

The rostral end of the third ventricle is formed by the organum vasculosum laminae terminalis (ovlt) (Fig. 23, section 9). It is composed of flat ependymal cells with round and oval nuclei and a dense capillary network. Not all the ependymal cells reach up to the ventricular lumen, some have basal processes. Immunocytochemical, biochemical and radioimmunoassays show that the ovlt is rich in biogenic amines and neuropeptides. The dopamine concentration is especially high, but it contains also other catecholamines and their synthetising enzymes (Saavedra et al., 1976). With biochemical methods an average quantity of serotonin (Saavedra et al., 1976) and tryptophan hydroxylase (Brownstein et al., 1975) was detected in the ovlt. The relatively high choline acetyltransferase activity (Saavedra et al., 1976) is in fair agreement with the histochemical observation demonstrating a strong acetylcholinesterase activity in the ovlt (Lewis and Shute, 1967; Mészáros et al., 1969).

Numerous immunocytochemical examinations have unanimously shown that the ovlt is fairly rich in luteinizing hormone-releasing hormone (LH–RH) (Barry et al., 1973; Zimmerman et al., 1974; Kozlowski and Zimmerman, 1974; Baker et al., 1975; Barry and Dubois, 1976; Kozlowski and Frenk, 1977). This fact was corroborated also by the radioimmunoassays, according to which, after the median eminence, in the brain the ovlt contains LH–RH in the highest concentration (Brownstein et al., 1975; Kizer et al., 1976). Since the organ does not have LH–RH-containing cells, it is assumed that the axons of the LH–RH neurons, situated in the adjacent medial preoptic nucleus, terminate here (Barry and Carette, 1975; Hoffman, 1976; Barry, 1977; Silverman et al., 1977). It should be noted that there is a direct neural, and a rather close vascular connection between the medial preoptic nucleus and the ovlt (Palkovits et al., 1978).

Apart from LH–RH, thyrotropin-releasing hormone (Kizer et al., 1976), somatostatin (Palkovits et al., 1976; Weindl and Sofroniew, 1978), and oxytocin (Zimmerman and Antunes, 1976) may be revealed in the ovlt. Most probably, the nerve terminals contain these substances, as with immunocytochemical methods only neurophysin-containing perikarya could be demonstrated (McNeill, 1977).

Caudal to the ovlt, the supraoptic recess can be observed (Fig. 23, sections 10–11). Its floor is formed by markedly flattened cells with elongated oval nuclei that are parallel with the ventricular floor. The side walls of the supraoptic recess are formed of cuboidal cells with round nuclei, arranged in one layer (Fig. 24).

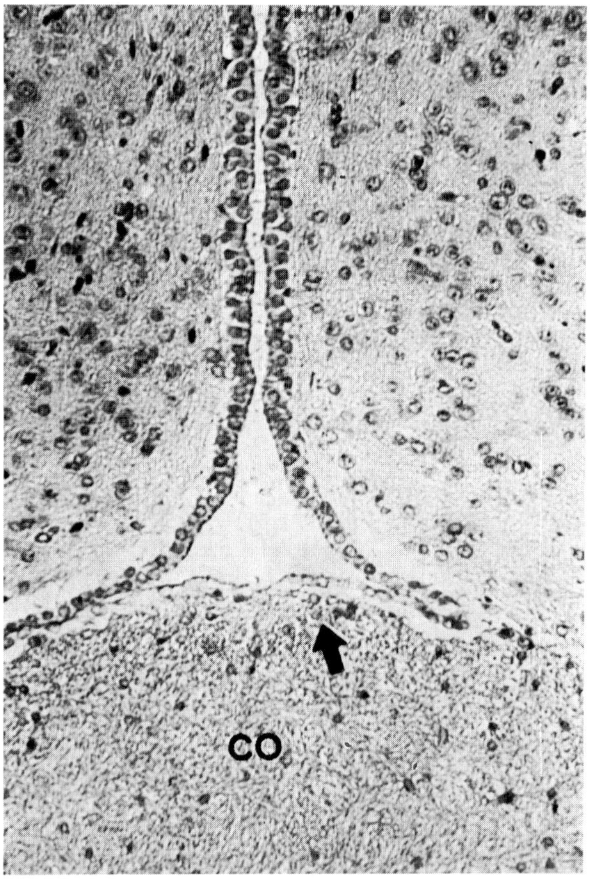

Fig. 24. Supraoptic recess at the bregma level. Frontal section, haematoxylin–eosin staining. The flat ependyma on the floor of the recess (→), immediately above the optic chiasma (CO), can be clearly distinguished from the cuboidal ependyma of the side wall. Magnification: ×220

Under the anterior commissure there are flocks of ependymal cells. The cell nuclei are round (Fig. 25A). The side walls of this segment are formed by monolayered cuboidal ependyma, like caudally, in the so-called retrochiasmatic area (Fig. 25B).

The monolayered cuboidal ependyma also covers the side walls ventral and caudal to the interventricular foramen. The cells have round nuclei with cilia on their surface.

The side walls of the ventricle portion in the anterior hypothalamic region are covered by regular cuboidal ependymal cells (Fig. 26A, B, C). Scattered subependymal cells are also present. They are oriented parallel with the apical surface of the cells. Supraependymal cells are also found (Fig. 26A). The ependymal cells possess microvilli and cilia. The cell body contains mitochondria, smooth and rough endoplasmic reticulum, free ribosomes and a well-developed Golgi apparatus. The basal part of the cells is bordered by the neuropil (Fig. 26C).

Closely under the massa intermedia the ependyma is multilayered (Fig. 27A). Electron microscopy revealed also capillaries and ependymoglial cells between the ependymal cell. These cells are smaller than the ependymal ones. They have dark cytoplasms and round to oval, dark nuclei in contrast to the ependymal cells, which have large, light cytoplasms and round, light nuclei (Fig. 27B).

The apical surface of the ependymal cells of the third ventricle (except the tanycytes) is covered by cilia, microvilli and cell protrusions (Fig. 28A, B). The microvilli are distributed unevenly on the surface of the cells. Some of

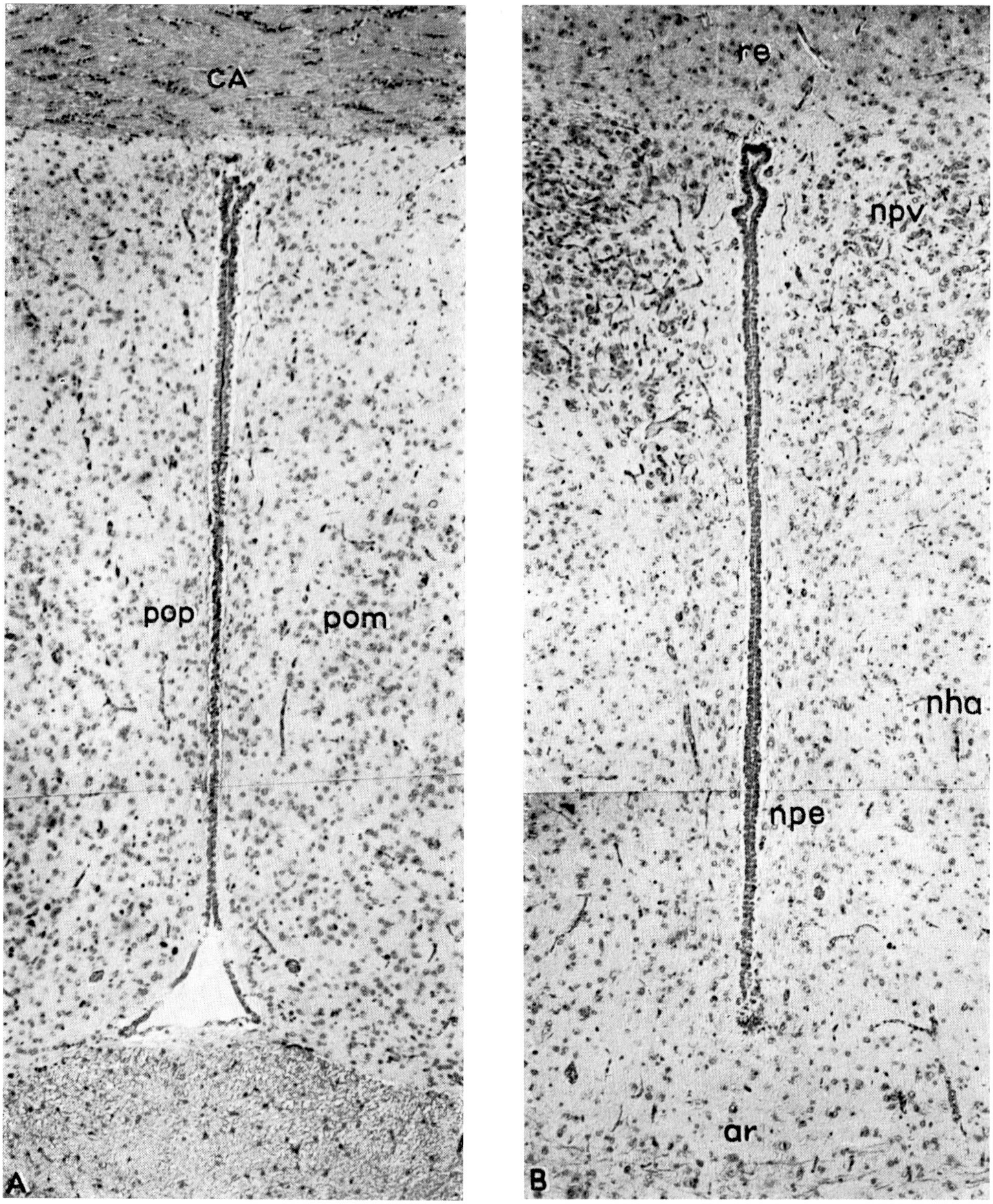

Fig. 25. Frontal sections of the third ventricle at the preoptic and anterior hypothalamic level. **A** — section at the plane of the bregma (0 mm). **B** — section at the level of the paraventricular nucleus (P1.8 mm). Magnification: ×100. For abbreviations see list

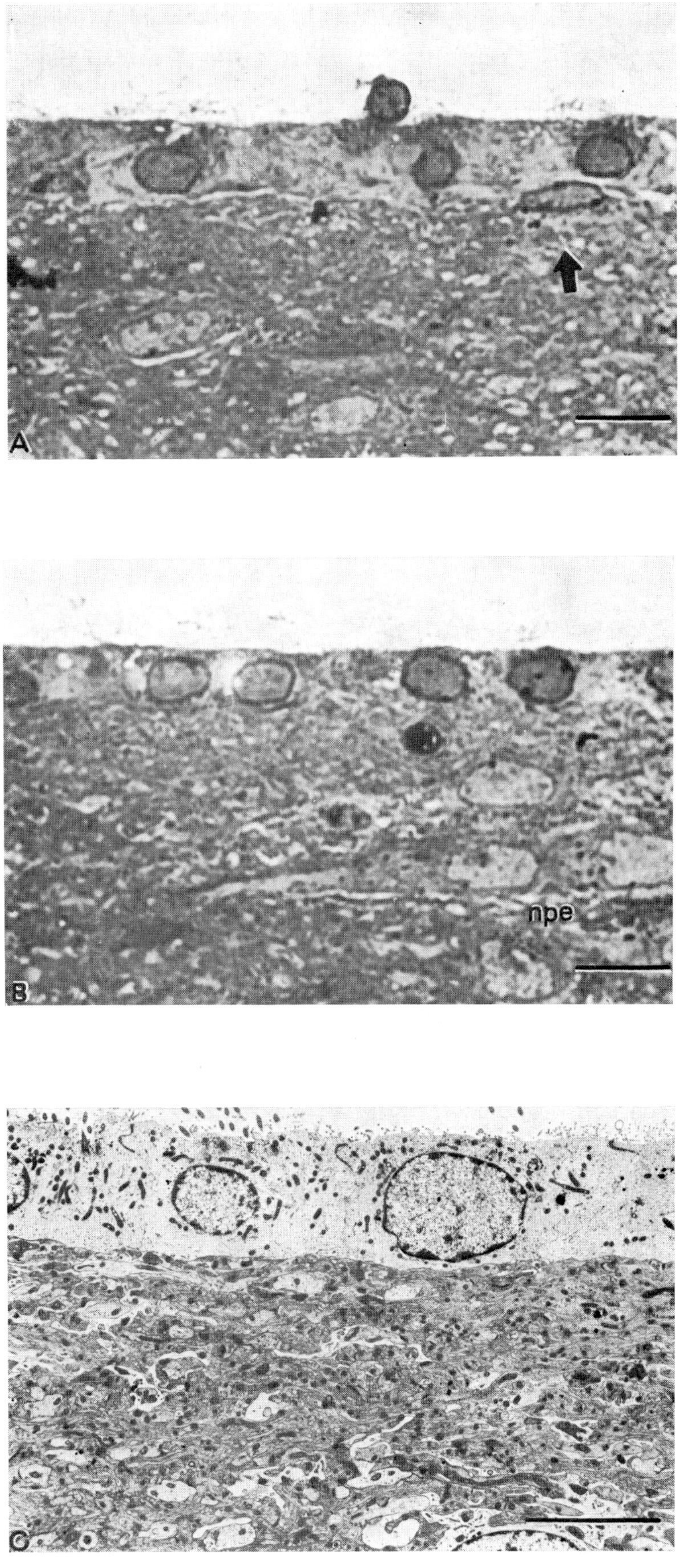

Fig. 26. Sections from the wall of the anterior hypothalamic ventricle. Regular cuboidal ependymal cells. **A** — occasional subependymal cells are visible (→); **B** — periventricular neurons (npe) are located close to the ependyma; **C** — electron micrograph of the cuboidal ependyma. Bar scales: **A** and **B** = 10 μm; **C** = 5 μm

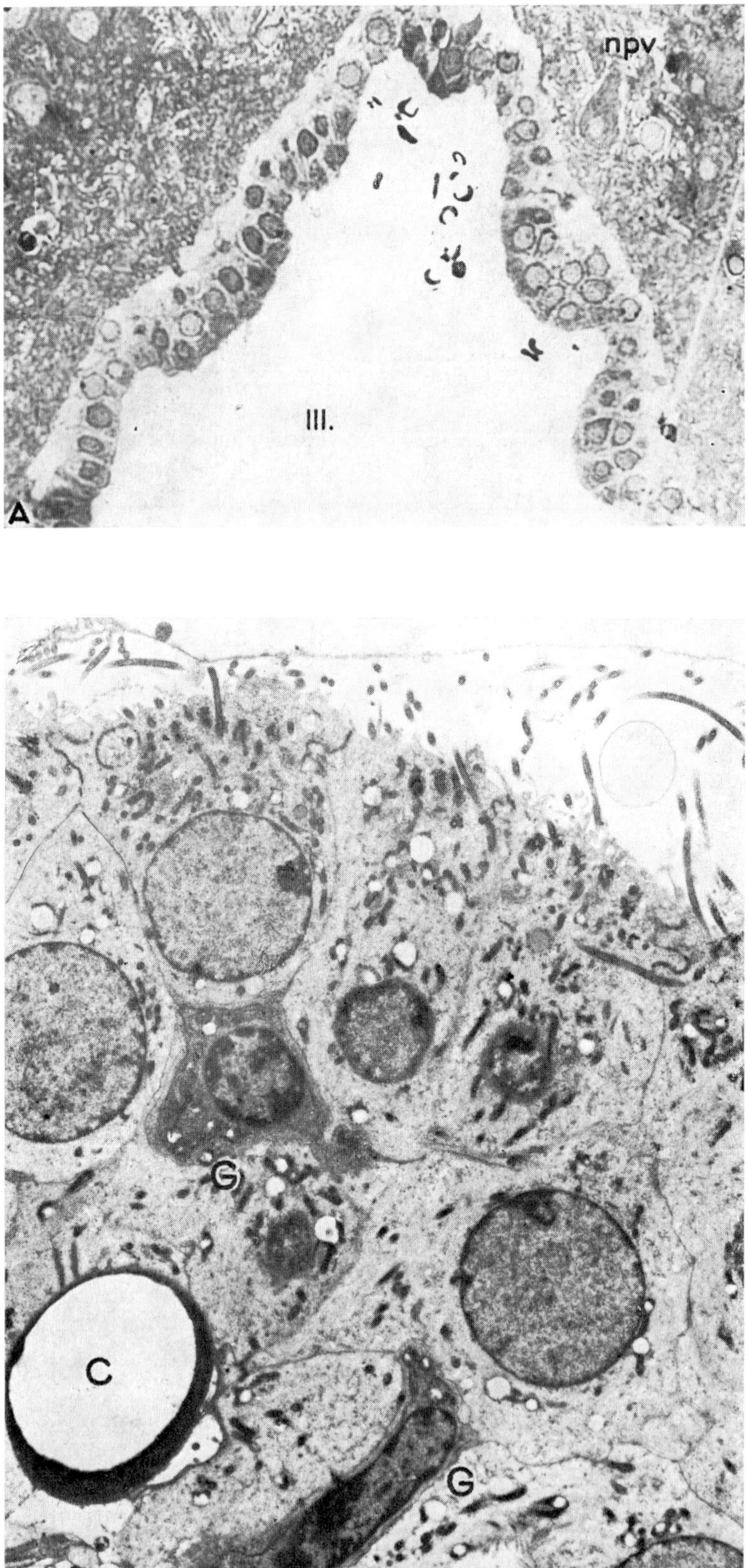

Fig. 27. Multilayered cuboidal ependyma from the top of the third ventricle (P1.5 mm to P2.1 mm). **A** — paraventricular nucleus (npv); III. — third ventricle. Magnification: ×400. **B** — glioependymal cells (G) and a capillary (C) within the multilayered ependyma. Bar scale = 5 μm

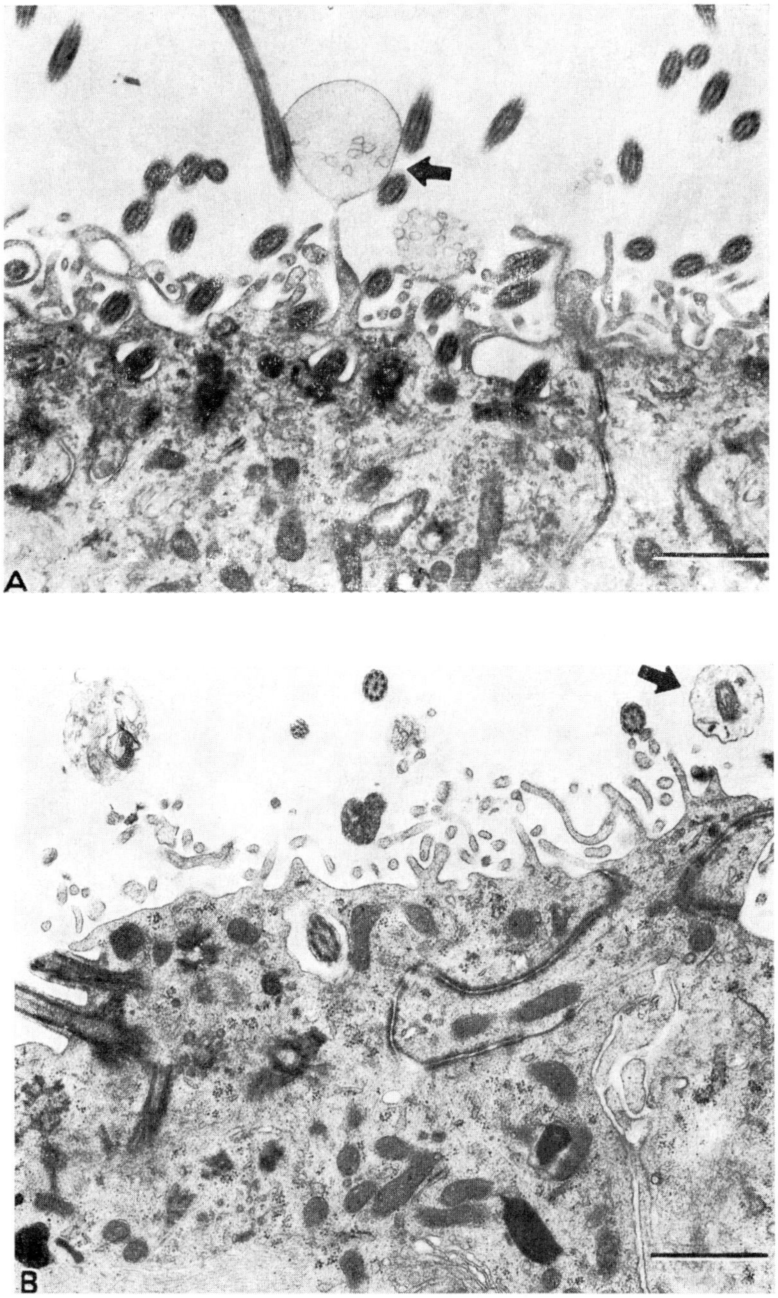

Fig. 28. Apical surface of the ciliated ependymal cell from the ventricle wall at the anterior hypothalamic level. **A** — supraependymal globules (→); **B** — intraventricular nerve fibres (→) can be detected on the apical surface. Bar scales = 1μ m

the microvilli form branches, while others form thickenings at their apical part.

The cilia usually have 9 + 2 pairs of microtubuli. Occasionally, we found cilia with cytoplasmic thickenings that contained vacuoles varying in size (Fig. 29A). Other cilia were surrounded by a cytoplasmic ring (Fig. 29B, as well as Fig. 28A). In the ventricle we noted unmyelinated intraventricular nerve fibres (Fig. 29C, D).

Closely under the surface of the cells there were rather large intracytoplasmic spaces (Fig. 30). In view of the presence of microvilli and cilia in these spaces we suppose their possible connection with the lumen of the ventricle. Contacts between the cells appear to be formed by the zonulae occludentes and zonulae adherentes (Fig. 30).

Intraventricularly, supraependymal cells with light cytoplasm and short processes can be observed. They generally contain large, dark cell nuclei (Fig. 31).

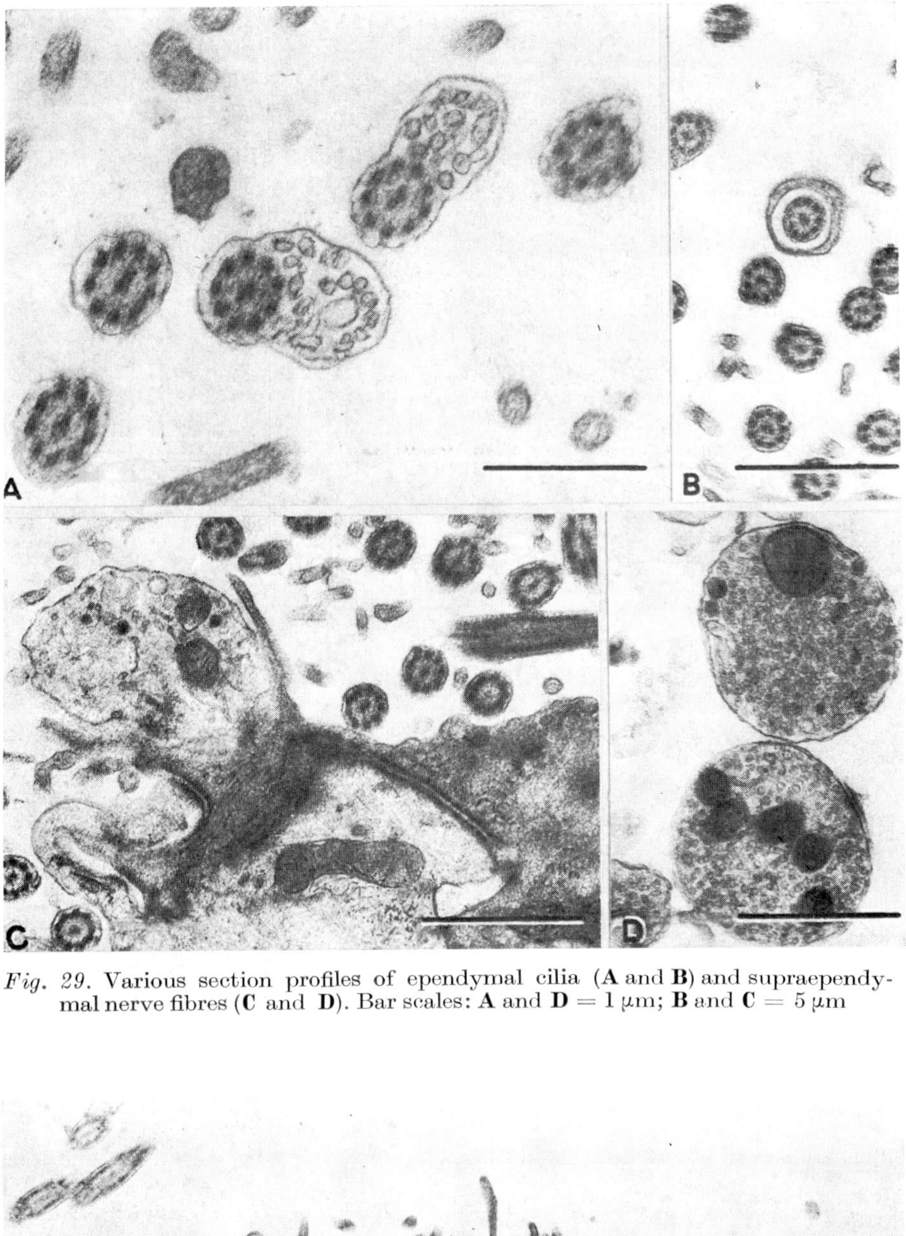

Fig. 29. Various section profiles of ependymal cilia (**A** and **B**) and supraependymal nerve fibres (**C** and **D**). Bar scales: **A** and **D** = 1 μm; **B** and **C** = 5 μm

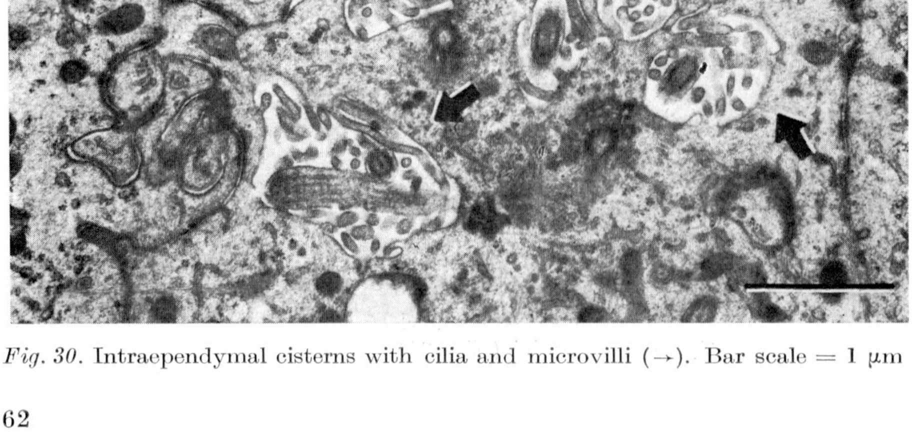

Fig. 30. Intraependymal cisterns with cilia and microvilli (→). Bar scale = 1 μm

Fig. 31. Supraependymal cells from the rostral portion of the anterior hypothalamus. Bar scale = 1 μm

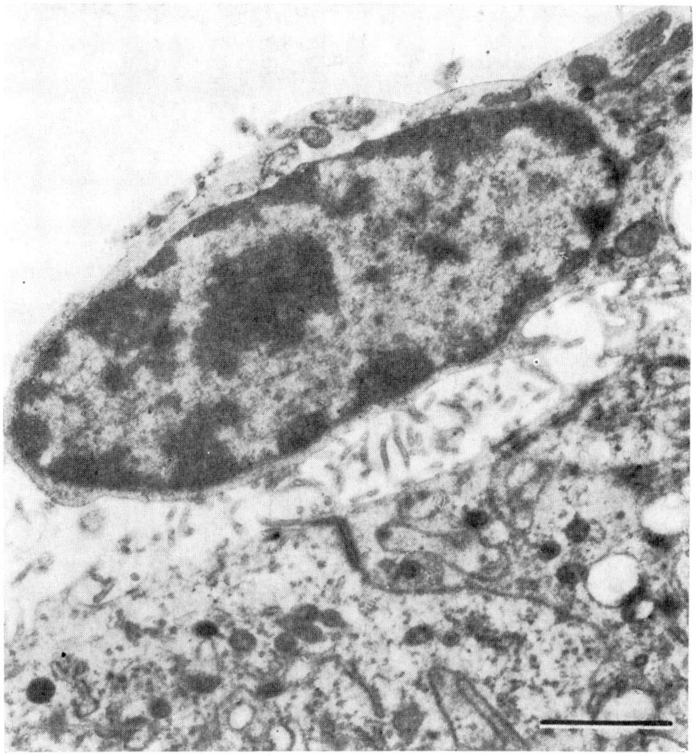

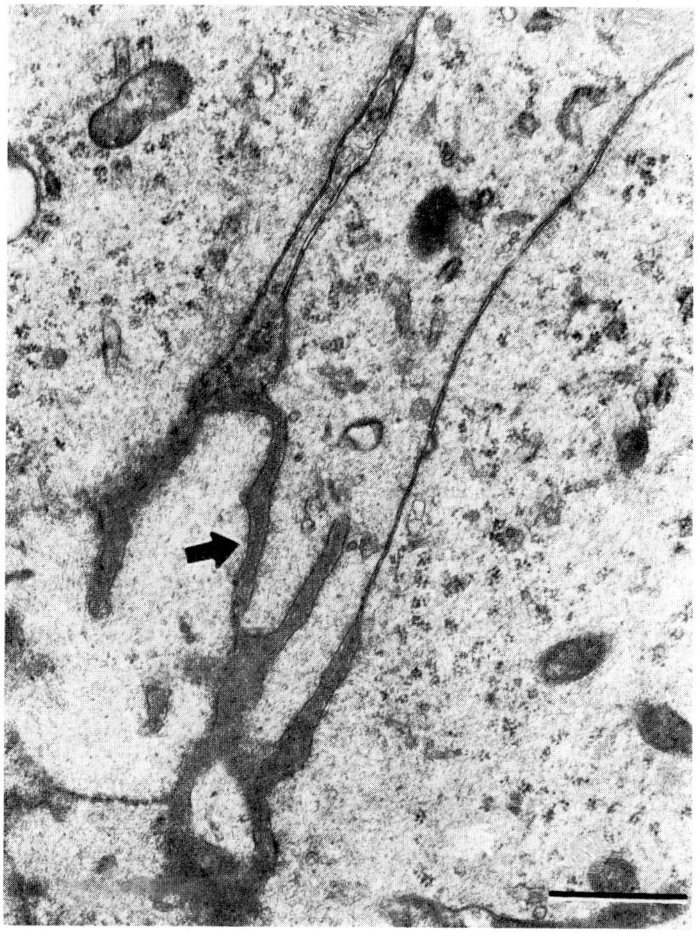

Fig. 32. Basal membrane labyrinths (→) under the ependymal layer. Bar scale = 1 μm

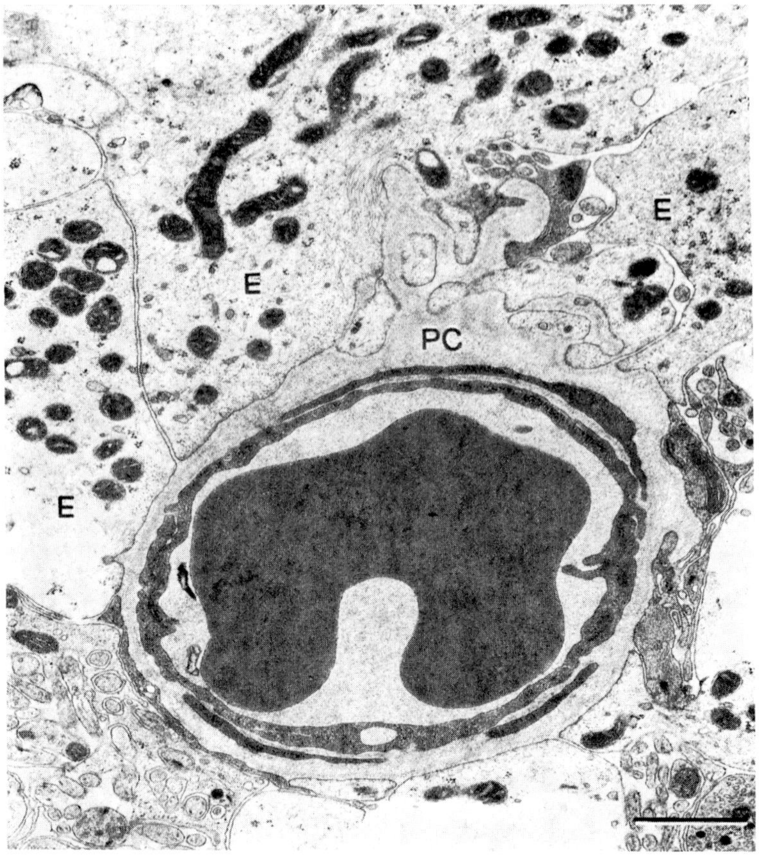

Fig. 33. Basal surface of ependymal cells (E) in topographical connection with a subependymal capillary. PC — pericapillary space. Bar scale = 1 μm

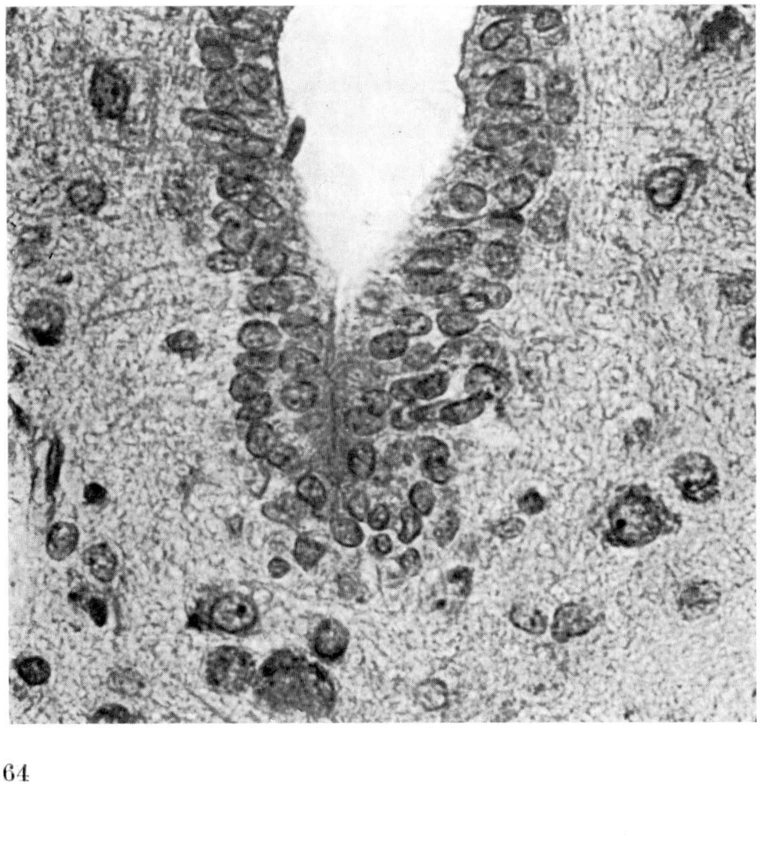

Fig. 34. Ventral tip of the third ventricle ("recess organ") at the rostral beginning of the median eminence (section 18 in Fig. 23, P2.1 mm) formed of multilayered ependyma. Magnification: ×640

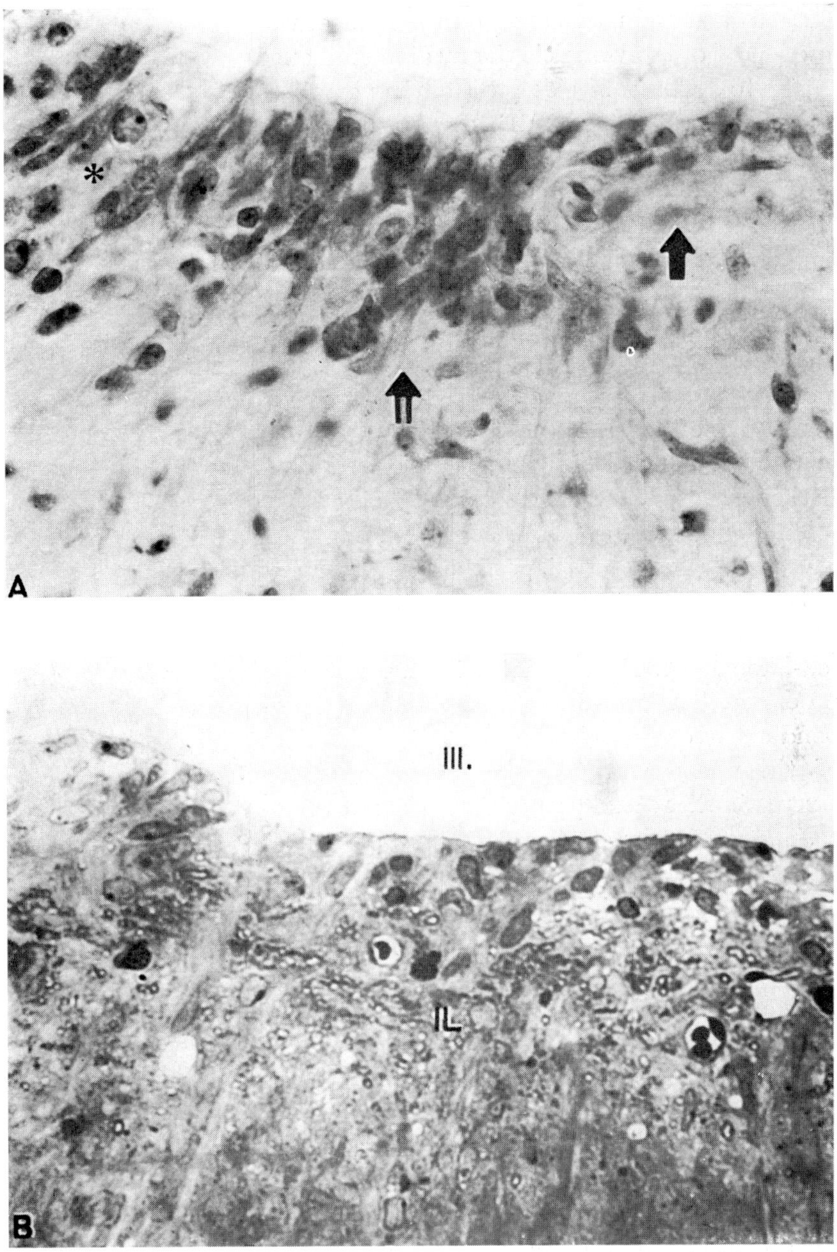

Fig. 35. The ependymal roof of the median eminence in paraffin (**A**) and semithin sections (**B**). III. third ventricle, IL — internal layer of the median eminence. Various types of ependyma can be distinguished: cuboidal to flat, monolayered ependyma in the middle portion of the median eminence (→), multilayered ependyma (⇒) ependymal cell groups are bordered by tanycytes (asterisk). Magnification: ×400

The basal part of the ependymal cells is either facing the basal membrane-like labyrinth (Fig. 32) or the pericapillary space (Fig. 33). In other places, the basal part of the ciliated ependyma is either adjacent to the surrounding neuropil (without any special subependymal parallel fibre) or to the subependymal glioependymal cells.

Behind the retrochiasmatic area, from P2.1 mm (Fig. 2B), there is the tuberal portion of the third ventricle.

The lower part of the ventricular lumen widens and forms the "recess organ" (Fig. 34). It is composed of multilayered ependyma with round to oval nuclei.

Backward from this area, on the floor of the lower part of the ventricle the median eminence begins to appear; the side walls of the ventricles are covered with tanycyte ependyma, i.e. the ependymal cells have long, ramifying basal processes. The surface of the tanycytes is devoid of cilia. The frontal section of this tuberal portion of the third ventricle is demonstrated in Fig. 36A. Various types of the ependymal regions can be distinguished such as the roof or the median eminence; arcuate tanycyte zone; monolayered tanycyte zone; irregular tanycyte zone; double-layered tanycyte zone; mixed zone; light cell zone with tanycytes; light cell zone without tanycytes; and cuboidal ependyma.

The ependyma on the floor of the median eminence is uneven. It is formed by ependymal cells with differently shaped nuclei. The third type is formed by ependymal cells with large, rod-like nuclei. Ependymal cells possess basal processes. The rod-like nuclei are either perpendicular or parallel to the apical surface of the ependyma. The ependyma is multilayered in the rostral part of the median eminence. The middle portion of the ventricular floor is lined with monolayered ependyma, and its lateral parts with multilayered ependyma. In the lateral sulcus (which is the corner between the floor and side walls in the lower part of the third ventricle) there are cells with elongated nuclei (Fig. 35A, B).

The side walls of the third ventricle above the median eminence are lined with various types of ependymal cells which show also various arrangement patterns.

The arcuate tanycyte zone is localized along the arcuate nucleus (Figs 36 and 37). The ventricular wall is lined here with a multilayered tanycyte ependyma (Fig. 38A). Between the ependymal cells nerve cells of the arcuate nucleus are inserted. On the surface of the tanycytes numerous oval formations with a dark central part can be demonstrated (Fig. 38B). The cytoplasms of the cells contain vacuoles with dark content.

Dorsally to the arcuate tanycyte zone the monolayered tanycyte zone is found (Fig. 36). The cell nuclei are generally elongated (Fig. 39). The cells possess long, basal, branched processes containing mitochondria (Fig. 40). The apical cell membrane is uneven; it has microvilli

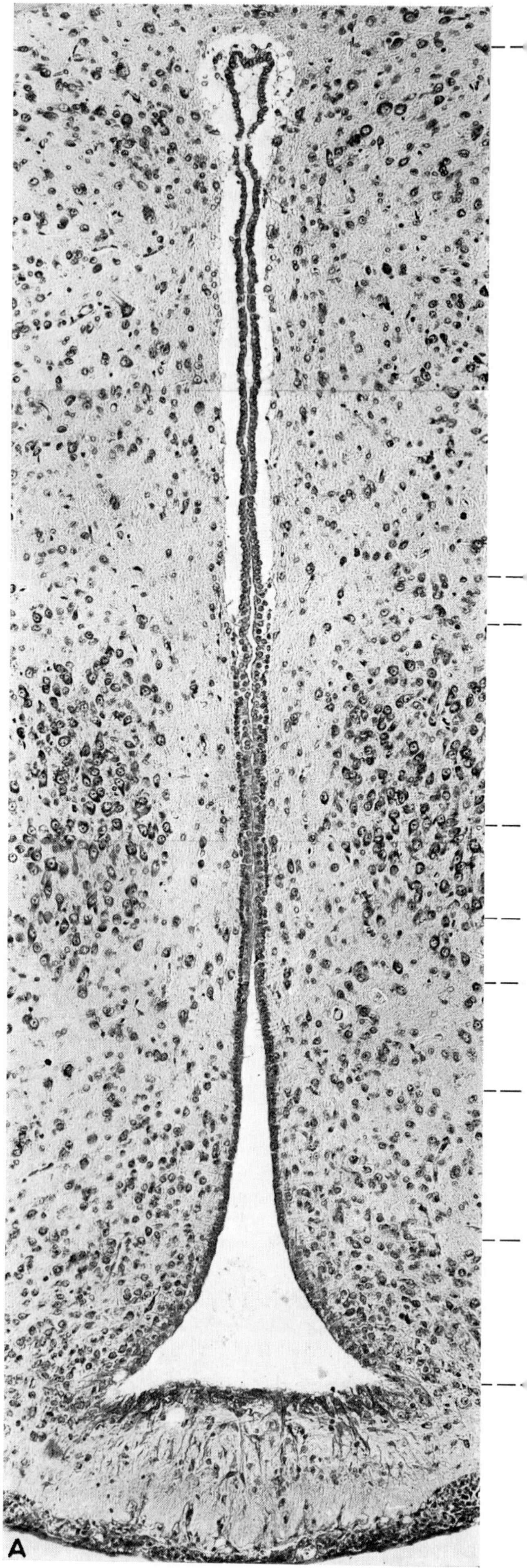

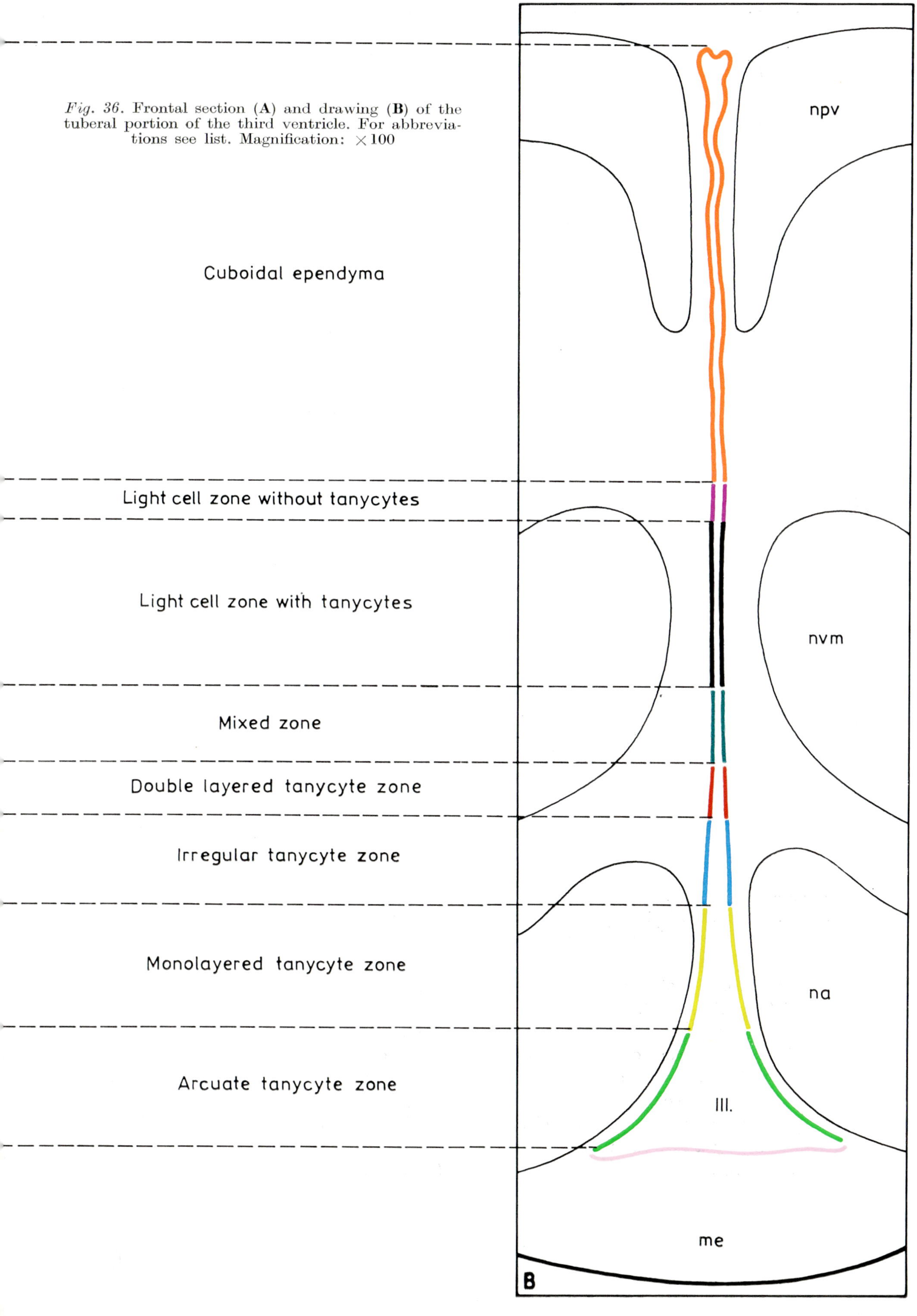

Fig. 36. Frontal section (**A**) and drawing (**B**) of the tuberal portion of the third ventricle. For abbreviations see list. Magnification: ×100

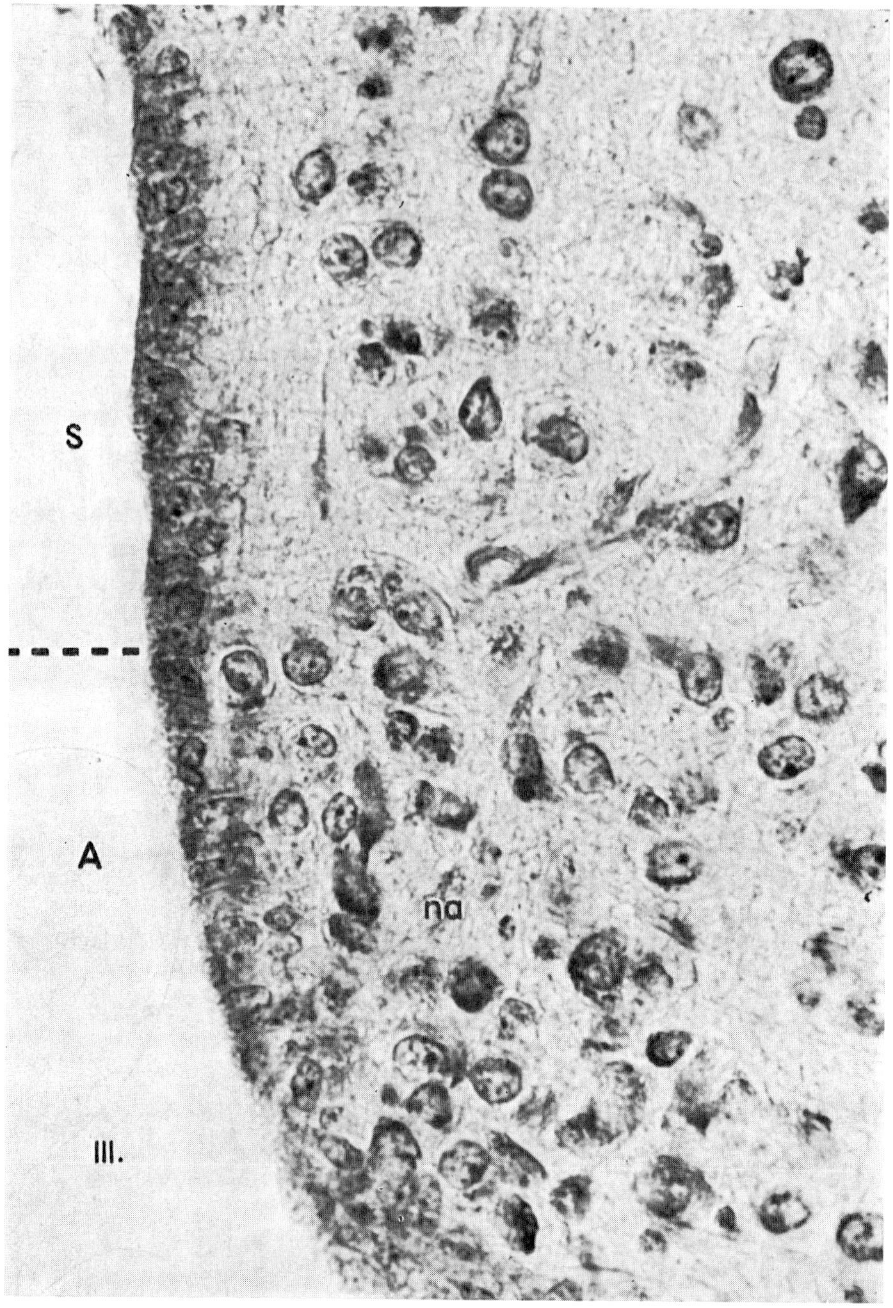

Fig. 37. Multilayered ependyma in the region of the arcuate nucleus (A) and "monolayered tanycyte ependyma" (S). na — arcuate nucleus; III. — third ventricle. Magnification: ×640

and round small cell protrusions with a dark central part (Fig. 41A, B, C). These formations are seen at various stages of maturation. The first stage is the bulging out of the apical cytoplasm. At later stages, this part of the cytoplasm becomes rounded and is connected with the surface of the cells only by a narrow peduncle. We suppose that finally these formations are released into the CSF.

More dorsally, above the monolayered tanycyte zone we find the irregular tanycyte zone (Fig. 36). It consists of tanycytes whose nuclei are elongated and are localized at various distances from the apical surface of the cells (Fig. 42).

The irregular tanycyte zone is followed dorsalwards by the double-layered tanycyte zone (Fig. 36).

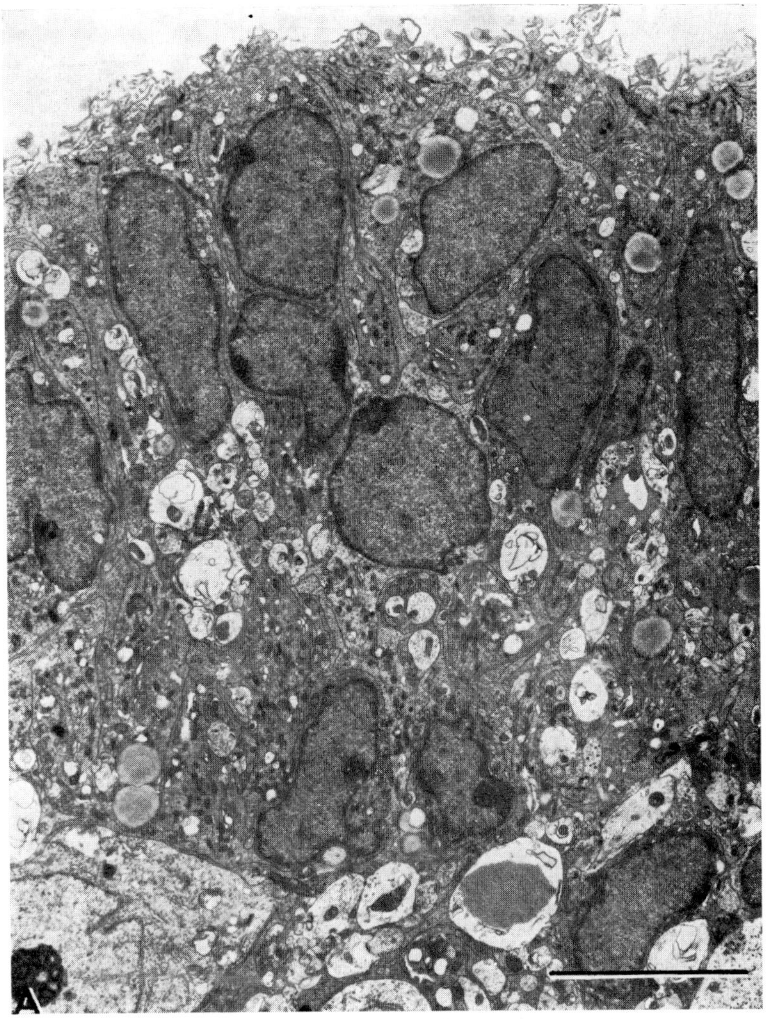

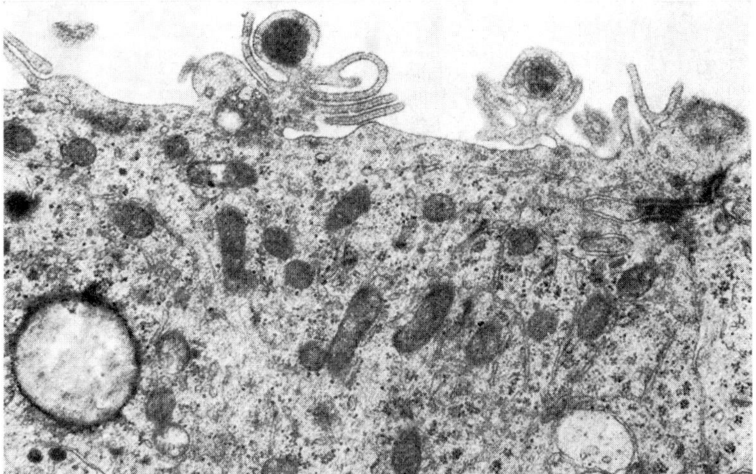

Fig. 38. **A** — "tanycyte ependyma" in the region of the arcuate nucleus. Bar scale = 5 μm. **B** — apical surface of tanycytes. Bar scale = 1 μm

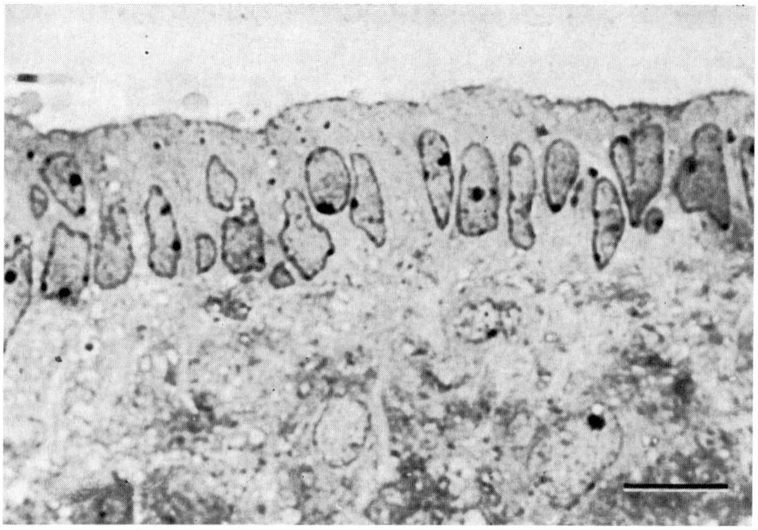

Fig. 39. "Single straight layer" of the tanycyte ependyma. Topography can be seen in Figs 36 and 37. Bar scale = 10 μm

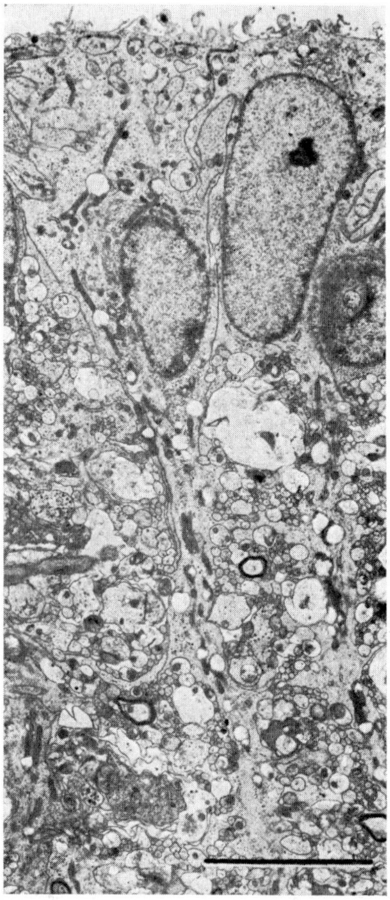

Fig. 40. Tanycyte perikaryon and basal processes. Bar scale = 5 μm

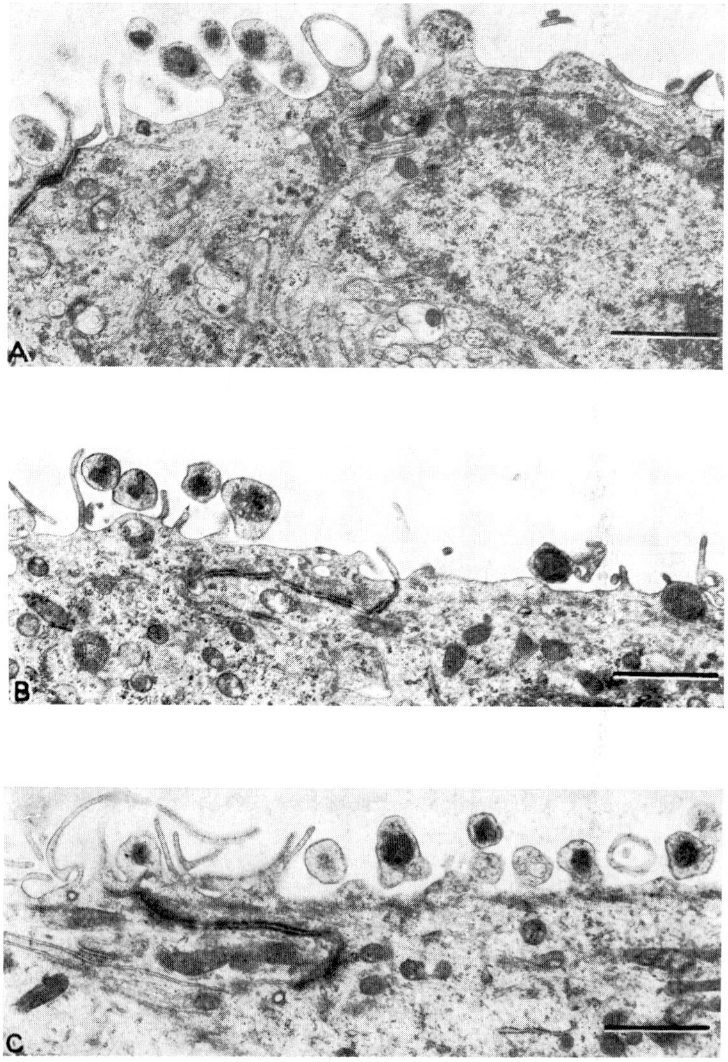

Fig. 41. **A**, **B**, and **C**. Various forms of appearance of the apical surface of tanycytes. Bar scale = 1 μm

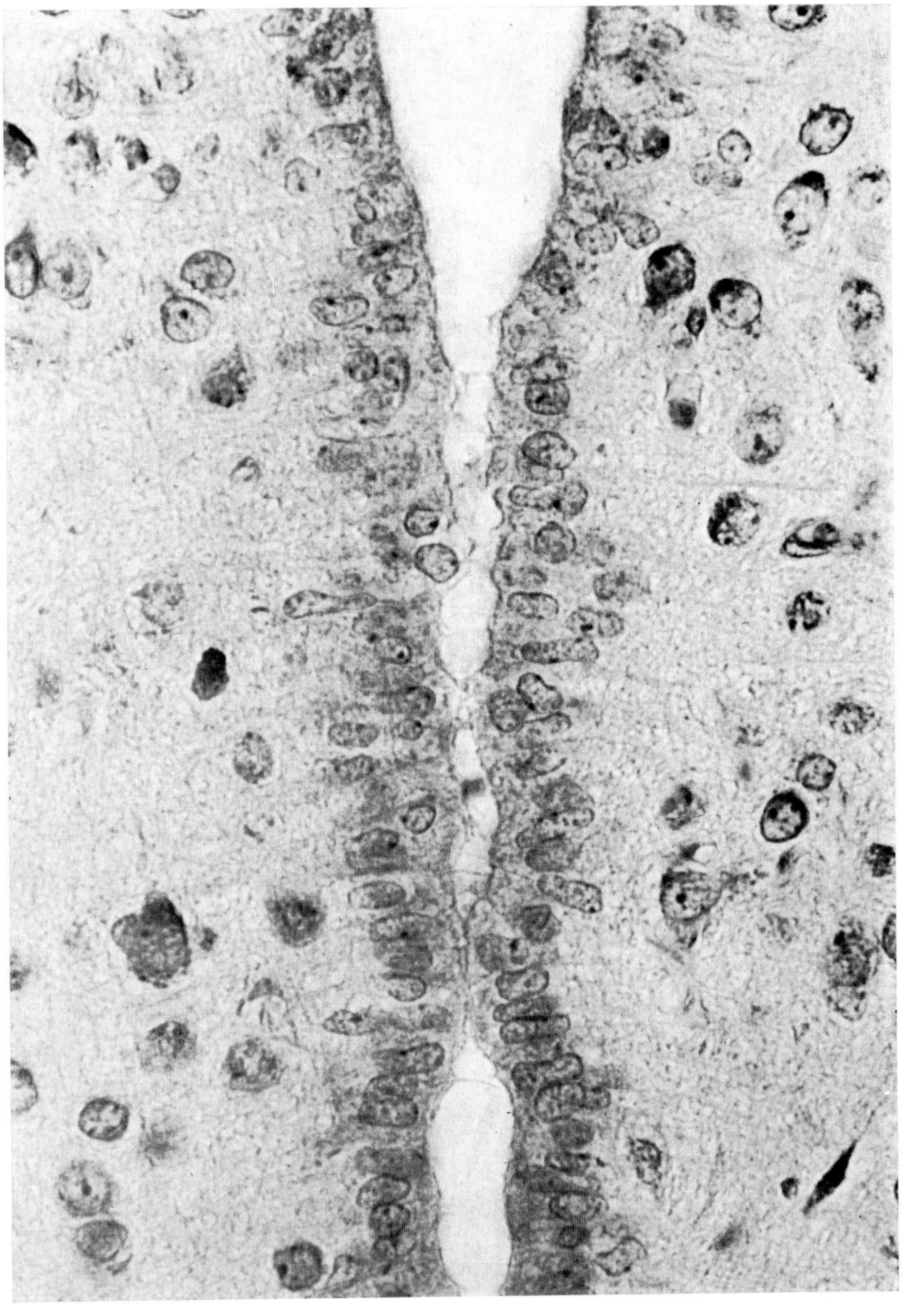

Fig. 42. Single, irregular tanycyte layer. Topography of this region can be seen in Fig. 36. Magnification: ×640

Figure 43 shows the double-layered tanycyte zone, below and above the mixed zone.

The apical surface of the ependyma is not undulated (Fig. 44).

The next zone dorsalwards (Fig. 36) is the mixed zone which consists of an uninterrupted outer layer of tanycytes (on the side away from the lumen) and an inner layer of cells with a large cytoplasm and light round nucleus (Fig. 45A). These latter do not form a continuous layer, and in some places they penetrate in between the tanycytes (Fig. 45B).

Above this area, the ependyma consists of two layers — an outer one formed by a single row of small cells with round nuclei, and an inner one (medially) formed by larger cells with a large cytoplasm and large, round nuclei (Fig. 46). The luminal surface is undulated. The cells are arranged in a ripple formation (Fig. 47A). This region is called the light cell zone with

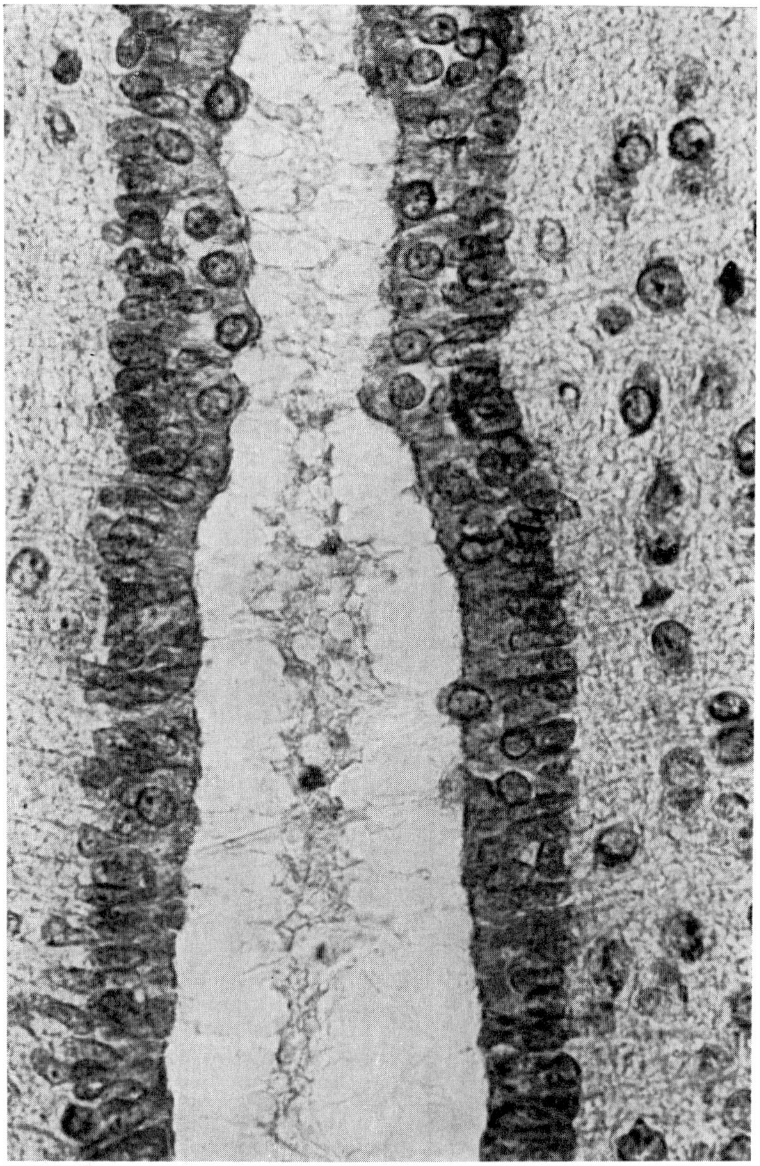

Fig. 43. Bilayered tanycyte ependyma with regular surface (lower two-thirds of the picture). In the upper third the regular ependyma is mixed with superficially localized "light cells" (mixed zone). Topography of these two regions can be seen in Fig. 36. Magnification: ×640

Fig. 44. Bilayered tanycytes with regular surface. Bar scale = 10 μm

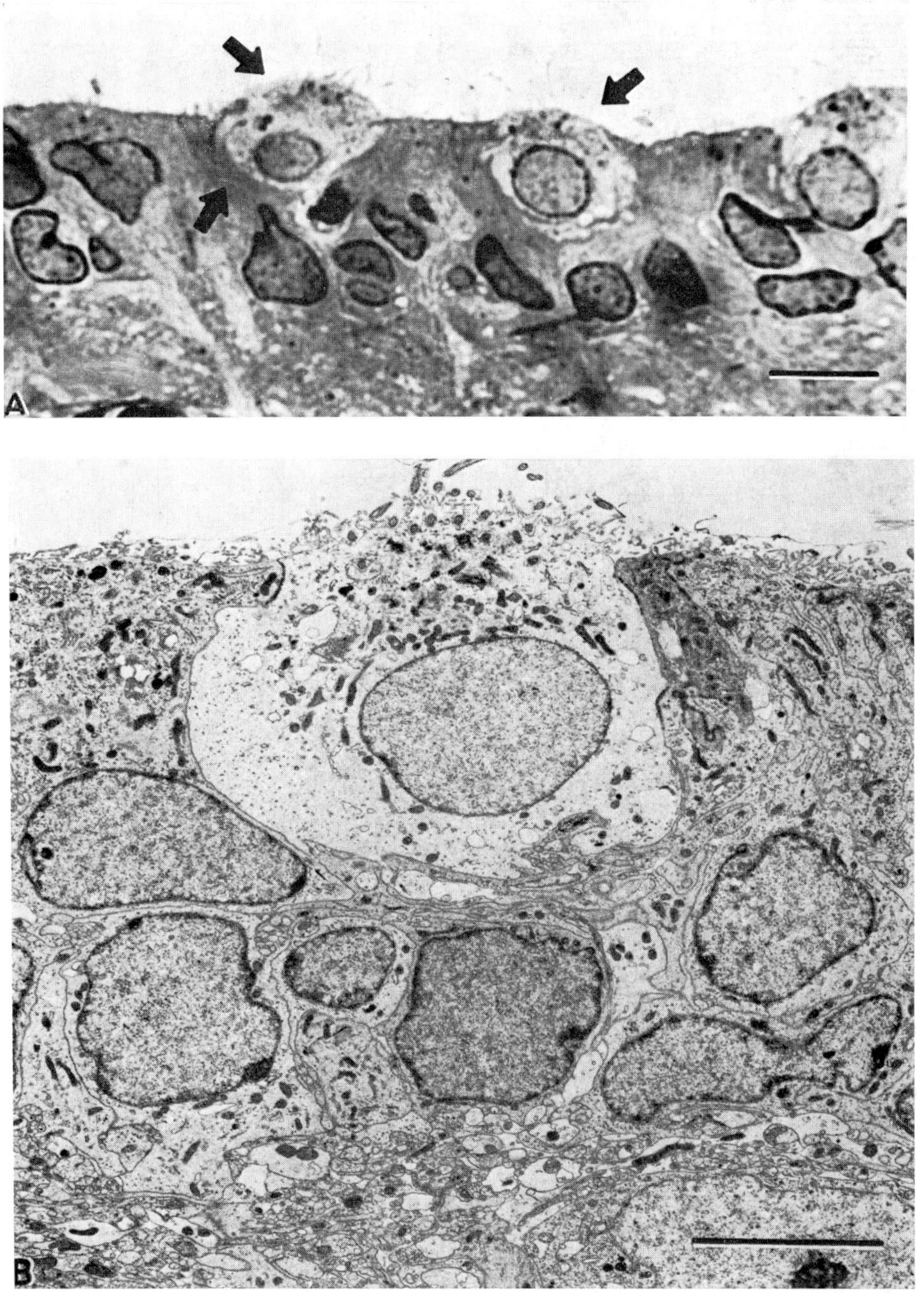

Fig. 45. Mixed zone of the tanycyte ependyma: tanycytes and "light cells". Topography can be seen in Figs 36 and 43; arrows in (**A**) point to the superficial "light cells". Bar scales: **A** = 10 µm; **B** = 5 µm

tanycytes (Fig. 36). By electron microscopy some tanycytes are observed which emit their processes to the ventricular lumen (Fig. 47B). Occasionally, these apical processes interrupt the layer of light cells and form a cavity with a large number of microvilli (Fig. 47C). Toward the roof of the third ventricle the ependyma becomes thinner and consists of a single layer of columnar cells (Fig. 48A, B). This is the light cell zone without tanycytes (Fig. 36). Upwards, the ependyma consists of cuboidal cells which form folds on the roof of the ventricle (Fig. 49). This area is the region of cuboidal ependyma. Monolayered ependyma is formed by cuboidal cells with (Fig. 50A) or without (Fig. 50B) subependymal cells.

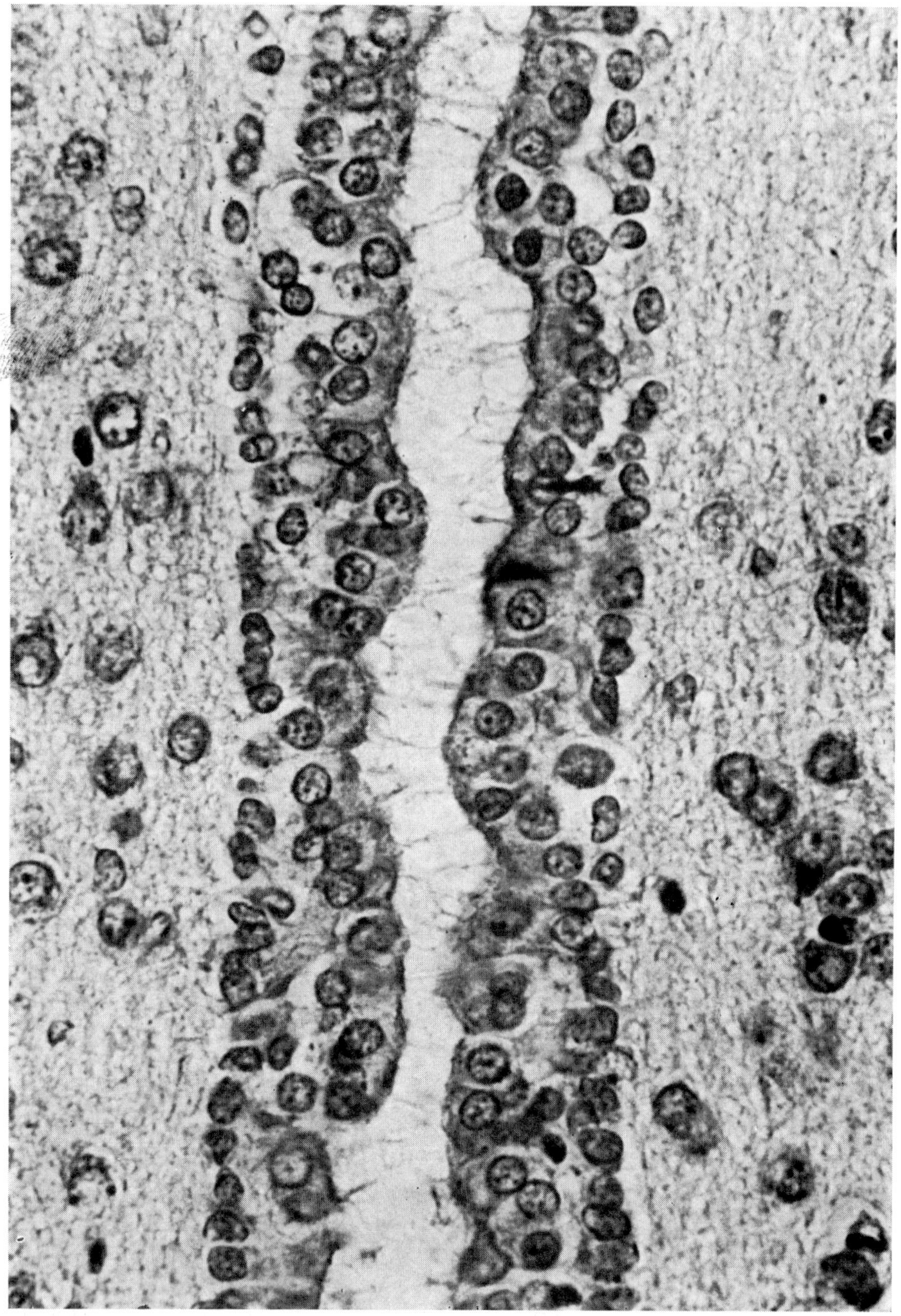

Fig. 46. "Light cell" zone with tanycytes. Topography can be seen in Fig. 36. The irregular apical surface of this zone is clearly demonstrable. Magnification: ×640

In coronal sections the inframamillary recess and the "connecting part" of the third ventricle occur together (Fig. 51A, B). The connecting part connects the ventral part of the third ventricle with the cerebral aqueduct. It is lined with cuboidal and columnar ependymal cells (Fig. 52). The ependymal cells have cilia. The cell nuclei are round or oval.

The inframamillary recess has different forms in frontal sections at the level of its rostral (Fig. 53) and caudal part (Fig. 57). The roof of the inframamillary recess is formed of mono- to multilayered ependyma creating large folds (Fig. 54A). The cells have large round nuclei and light cytoplasm. On the apical surface of the cell large cytoplasmic globules are visible both light

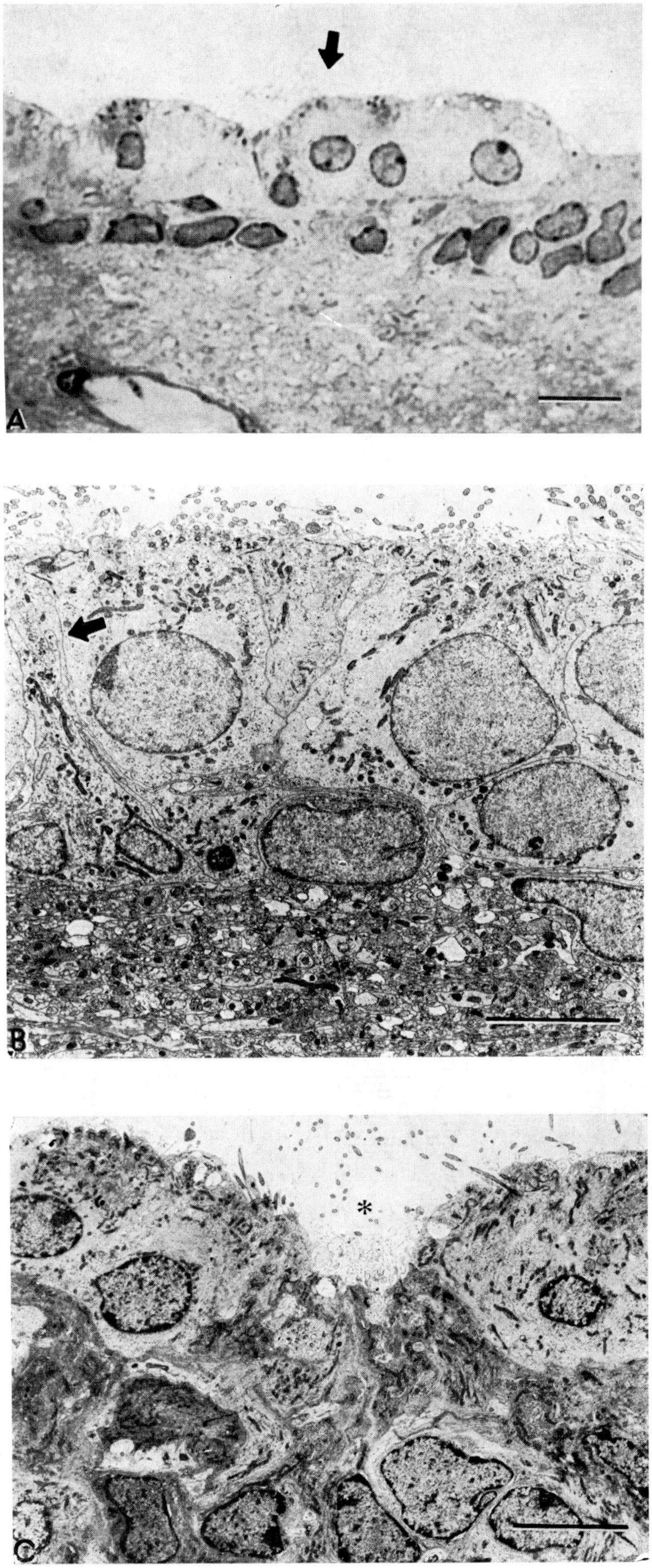

Fig. 47. "Light cell" zone with tanycytes. **A** — light cells (→); **B** — light cells dislocated by apical processes of the tanycytes localized under the light cells (→); **C** — asterisk indicates apical tanycyte processes. Bar scales: **A** = 10 μm; **B** and **C** = 5 μm

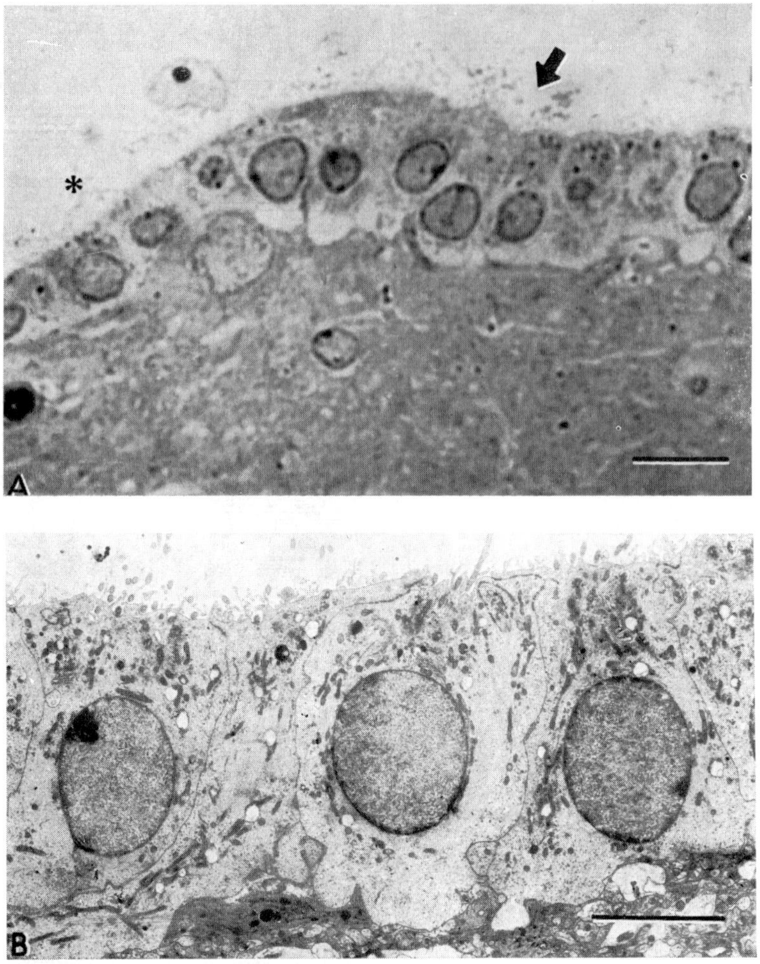

Fig. 48. **A** — "light cell" zone without tanycytes (→) and transition to cuboidal ependyma (asterisk). Topography can be seen in Fig. 36. Bar scale = 10 μm, **B** — "light cell" zone without tanycytes. Bar scales: **A** = 10 μm; **B** = 5 μm

(Fig. 54A) and electron microscopically (Fig. 54B, C). These globules have a well-preserved narrow stalk.

Under the roof, the side wall of the inframamillary recess is covered by multilayered tanycyte ependyma (Fig. 55A, B). The lower portion of the side wall is particularly thick, displaying 3–4 layers of overlying nuclei (Fig. 56A, B). Backwards, the lumen of the recess diminishes (Fig. 57). The roof ependyma becomes thinner, and in some places it is formed only by a single layer of cuboidal ependymal cells with round nuclei (Fig. 58). The side walls are covered with multilayered ependyma. The recess ends as a flock of ependymal cells.

Cell groups in the vicinity of the third ventricle

Although the ependyma covering the triangular recess as well as the surface of the interventricular foramen connecting the lateral and third ventricles are of the same type, the related subependymal structures are different. Fibres (commissura fornicis, stria medullaris, stria terminalis) and cell groups (paratenial, periventricular thalamic nuclei and bed nucleus of the stria terminalis) are covered by cuboidal ependyma (Fig. 2B/12–14). Being an exception, the subfornical organ is covered by a special type of ependyma.

There is no topographic correlation between the epithalamic ependyma and its related periventricular structures. The medial habenular and periventricular thalamic nuclei, and the stria medullaris are covered by flat or cuboidal ependyma (Fig. 2/B15–25).

The ependyma of the pineal recess also seems to be independent of its subependymal structures (Fig. 2B/26–31). At the beginning, the floor of this recess (Fig. 2B/26) is occupied by the supracommissural part of the subcommissural organ (Palkovits, 1965).

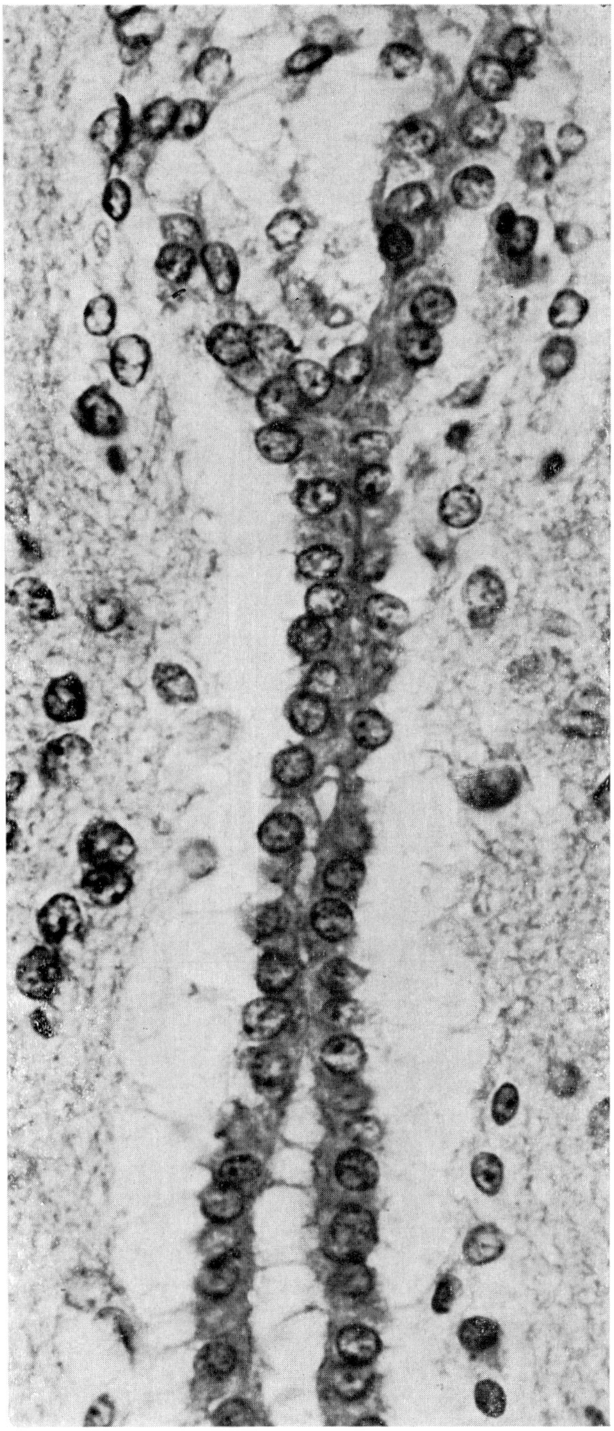

Fig. 49. Top of the third ventricle in the tuberal region with regular cuboidal ependyma cells. Topography can be seen in Fig. 36. Magnification: ×640

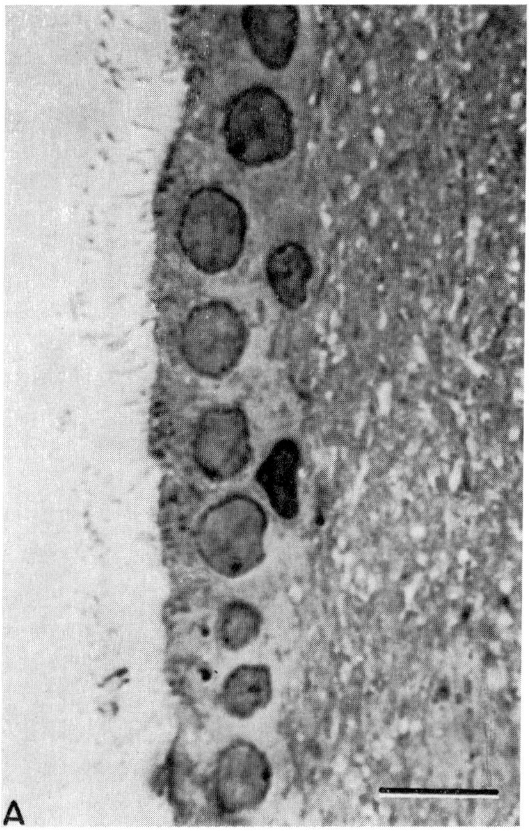

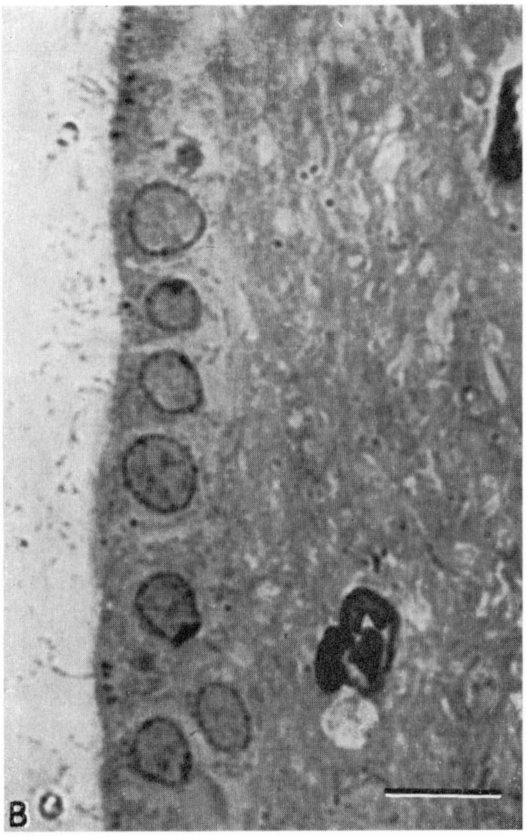

Fig. 50. Regular cuboidal cells from the top of the third ventricle with (**A**) and without (**B**) subependymal cells. Bar scales = 10 μm

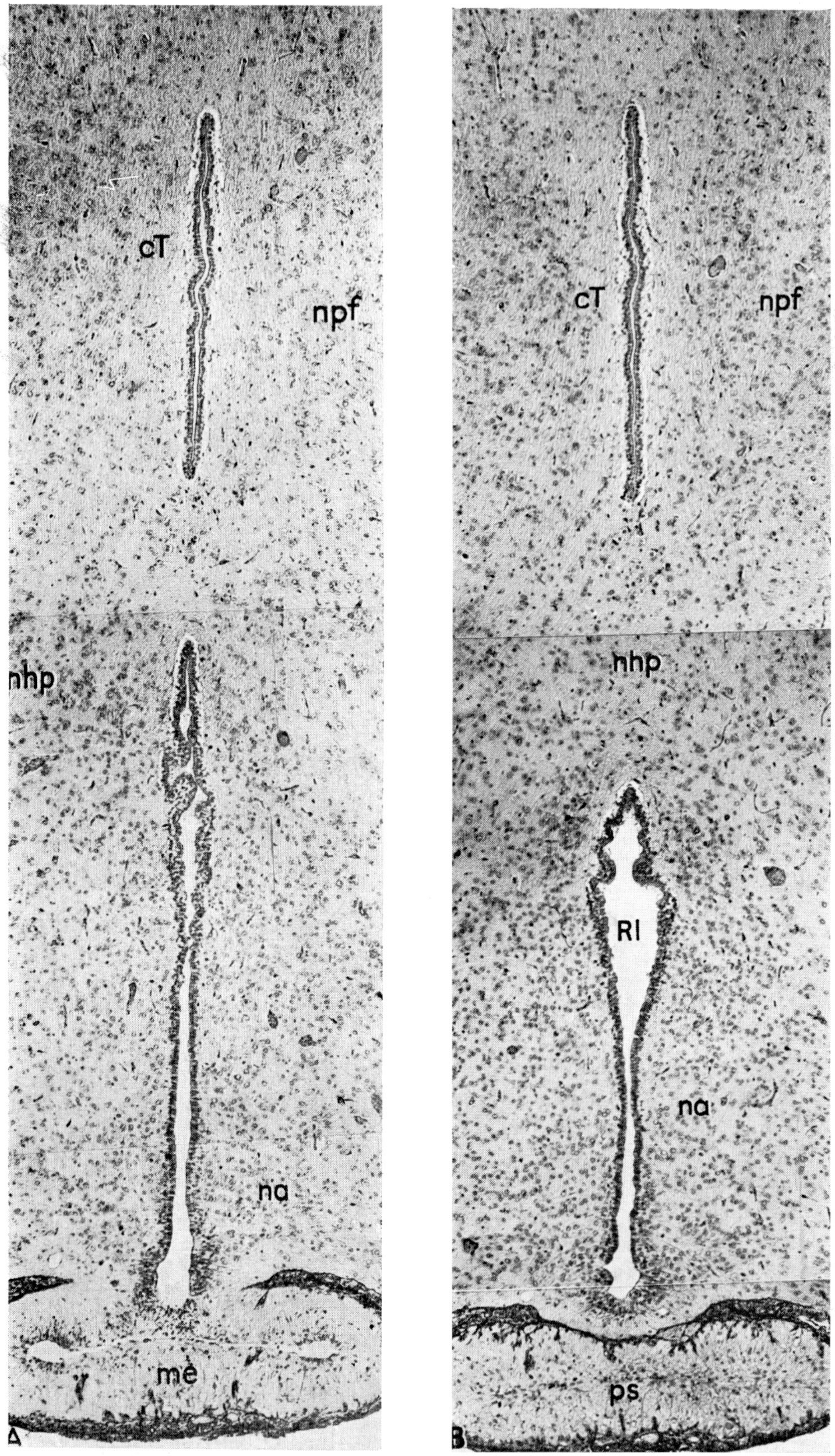

Fig. 51. Frontal sections of the third ventricle. The inframamillary recess and the connecting part between the third ventricle and cerebral aqueduct are seen at the posterior hypothalamic level (**A**—P3.6 mm; **B**—P3.8 mm). Magnification: ×100. For abbreviations see list

Fig. 52. Ependyma of the connecting part of the third ventricle formed by cuboidal and columnar cells. Bar scale = 10 μm

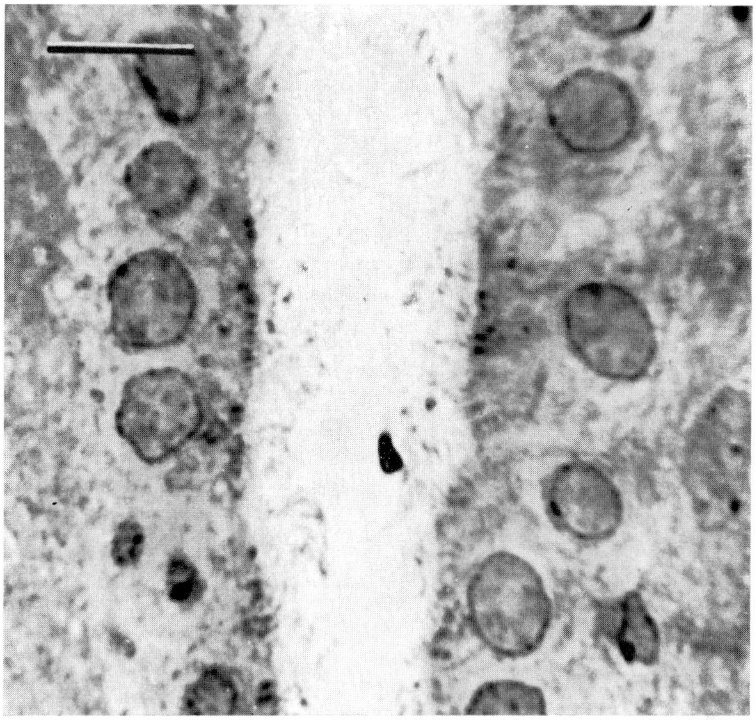

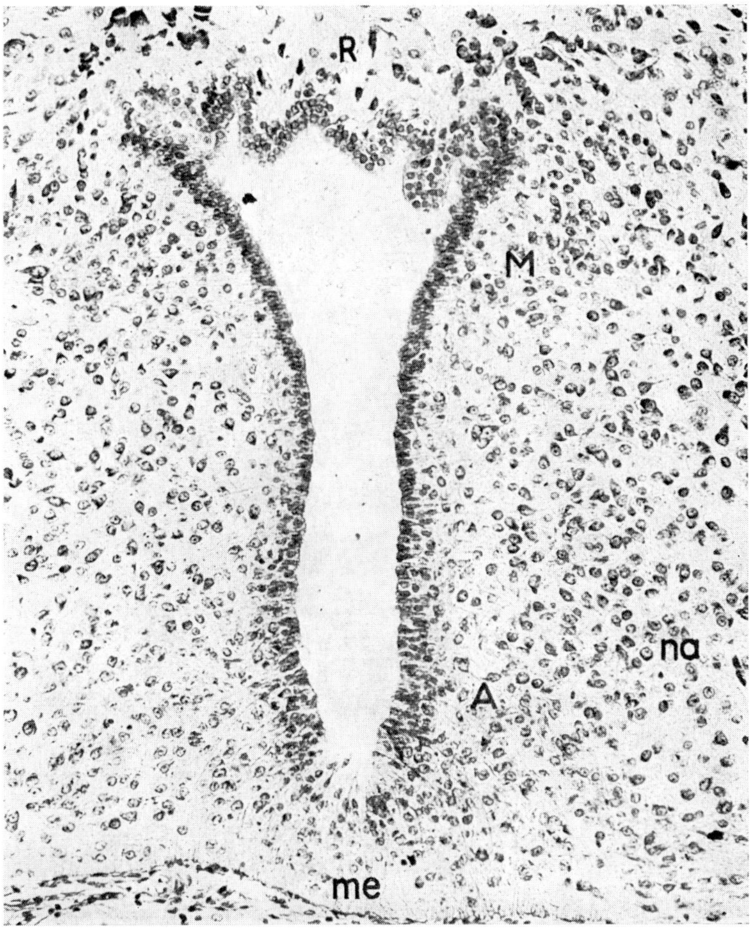

Fig. 53. Rostral part of the inframamillary recess (section 24, P3.9 mm). Three parts of the ependyma can be distinguished: 1. ependymal roof of the recess formed by "light cells" (R); 2. multilayered, regular tanycyte ependyma (M); 3. multilayered, "arcuate-type" tanycyte (A). na – arcuate nucleus, me – postinfundibular part of the median eminence. Magnification: ×200

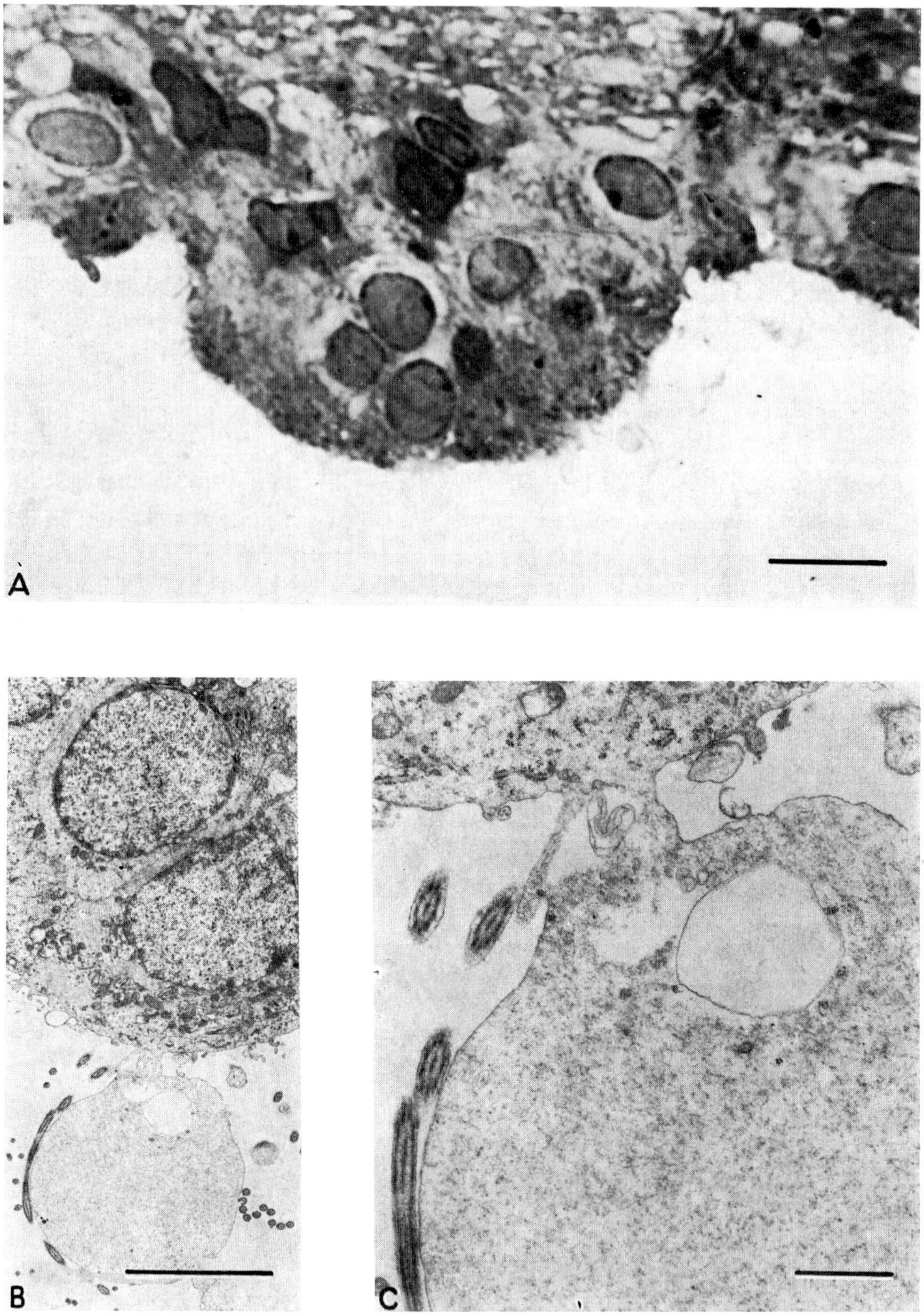

Fig. 54. Light cells with supraependymal protrusions (under high power in **B** and **C**) from the ependymal roof of the inframamillary recess. (Yellow area in Fig. 23, section 24; R in Fig. 53). Bar scales: **A** = 10 μm; **B** = 5 μm; and **C** = 1 μm

The organum vasculosum laminae terminalis is found at the rostral tip of the third ventricle (Fig. 2/9). The tanycyte-like cells of this organ constitute the lateral walls of the supraoptic recess, while the floor is covered by flat ependymal cells immediately on the dorsal surface of the optic chiasma.

The cuboidal ependyma on the anterior hypothalamic portion of the third ventricular wall is surrounded by the hypothalamic periventricular

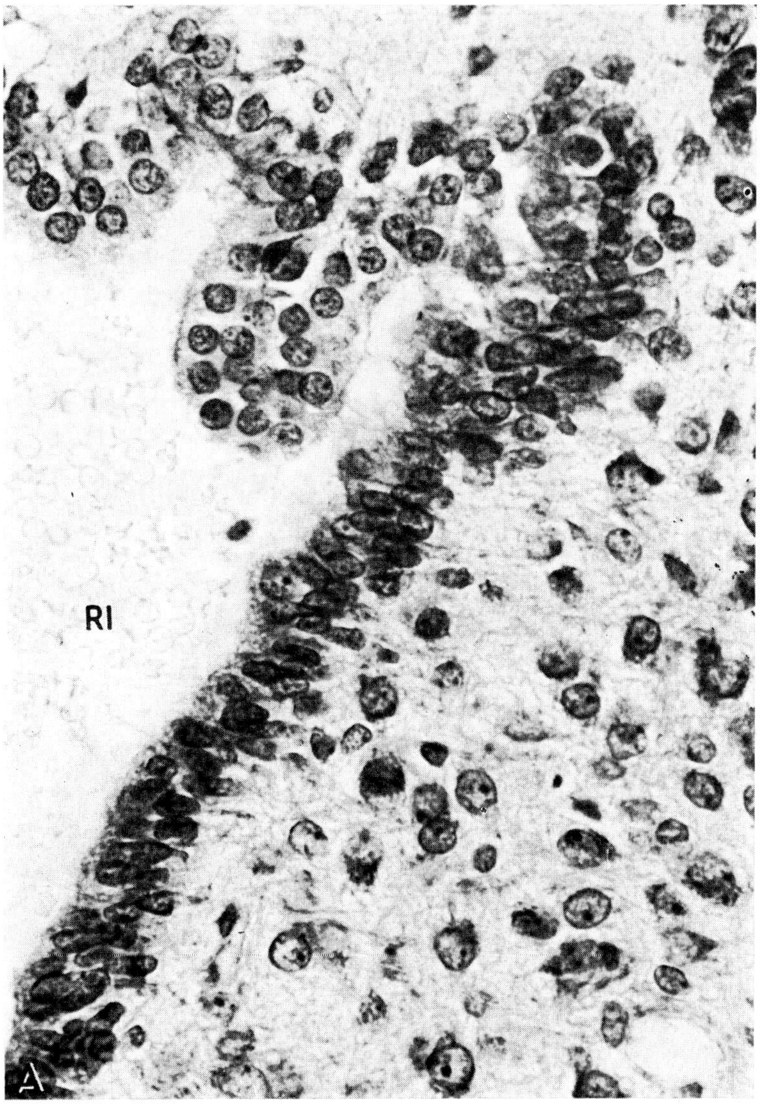

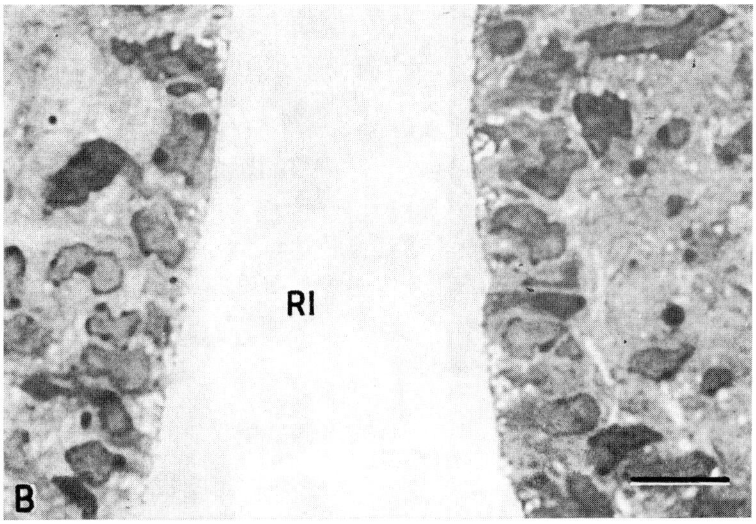

Fig. 55. Multilayered regular tanycyte ependyma on the upper part of the side walls of the inframamillary recess (RI). **A** — magnification: ×640, **B** — bar scale = 10 μm

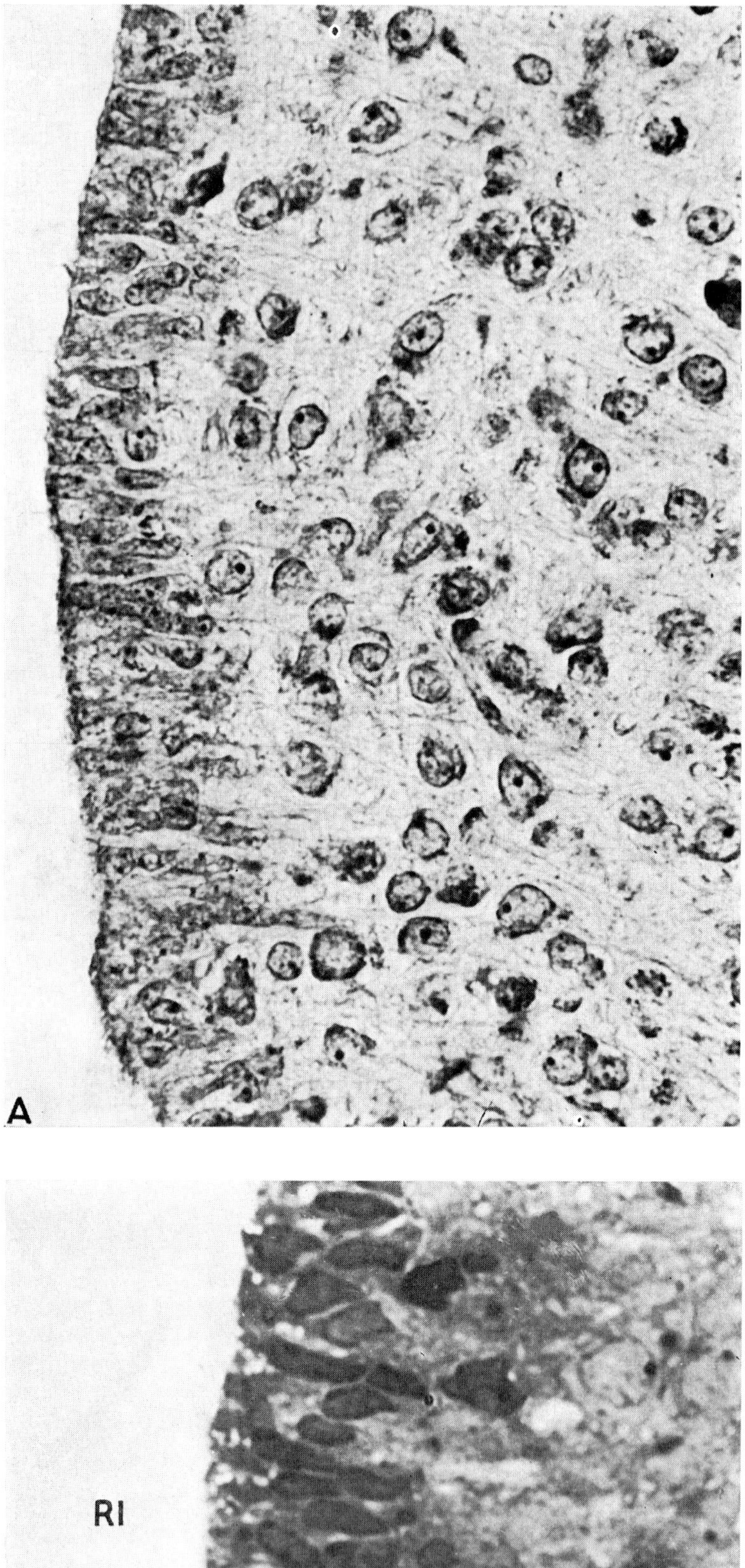

Fig. 56. Multilayered "arcuate-type" tanycyte ependyma on the lower part of the side wall of the inframamillary recess (RI). For topography see Fig. 53. The arcuate nucleus is infiltrated by numerous basal tanycyte processes. **A** — magnification: $\times 640$. **B** — bar scale = 10 µm

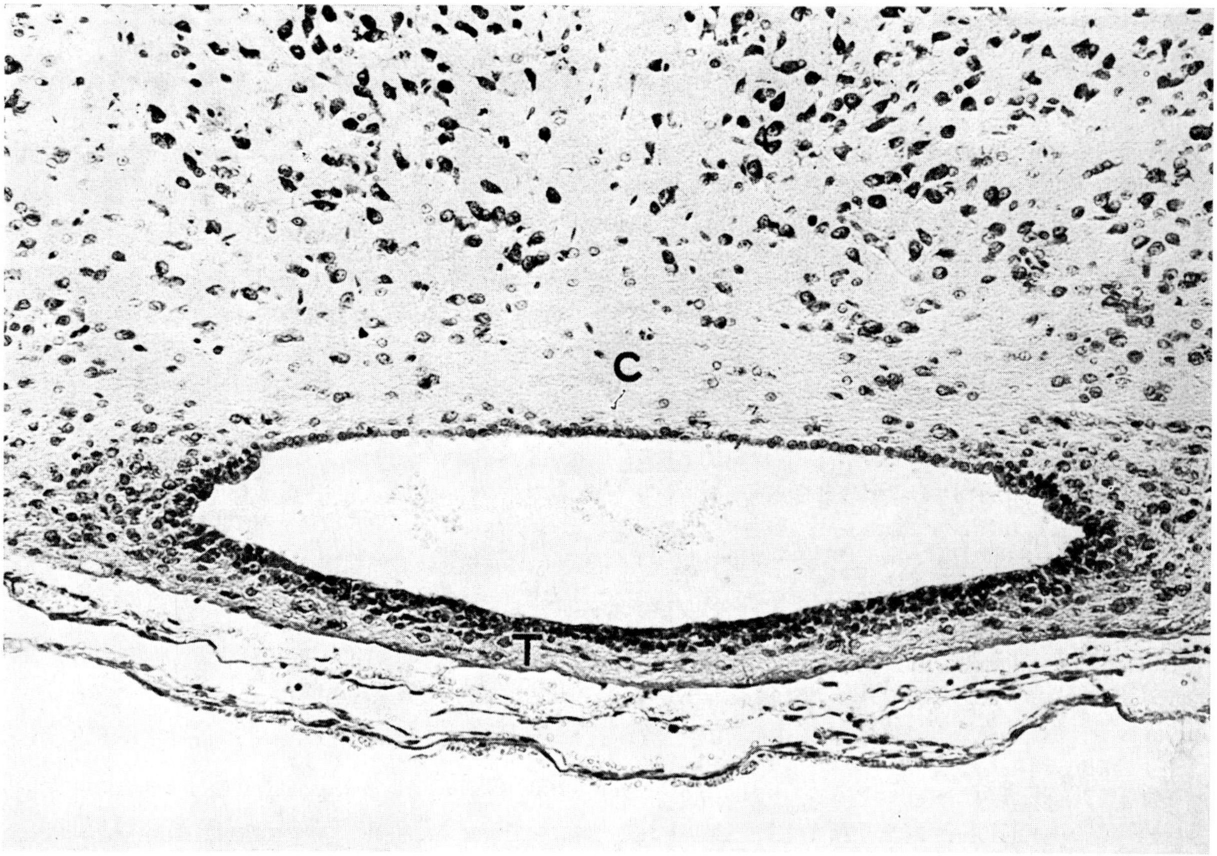

Fig. 57. Caudal part of the inframamillary recess (sections 25 and 26; P4.2 mm to P4.5 mm). Cuboidal ependyma on the roof of the recess (C) is indicated by green, the multilayered tanycyte ependyma on the floor of the recess (T) by blue in Fig. 23. Magnification: ×200

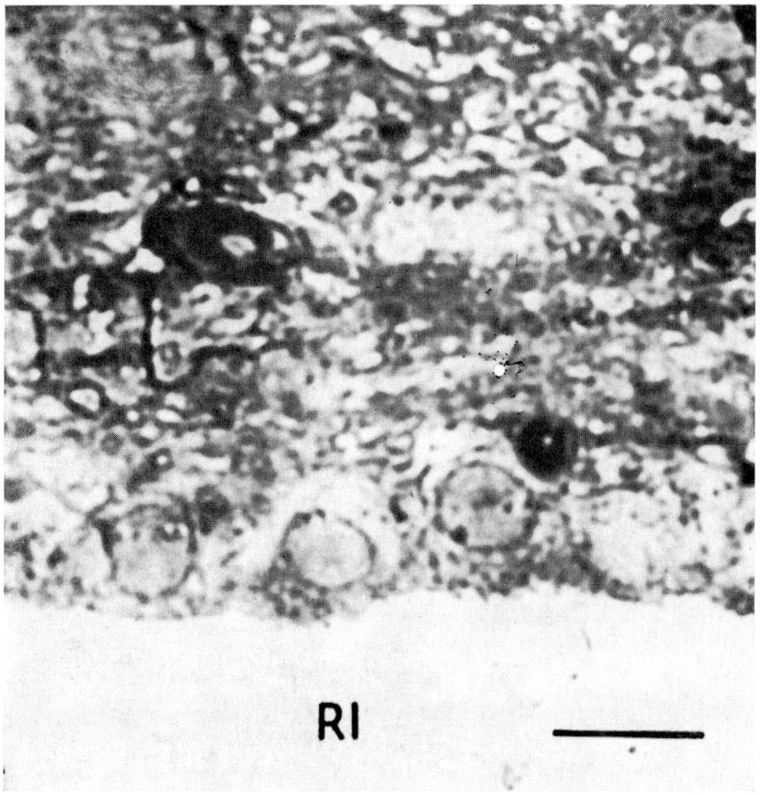

Fig. 58. Single layer of cuboidal ependyma from the roof of the inframamillary recess (RI). Bar scale = 10 μm

nucleus. In the rat, this nucleus consists of a few layers of cells and fibres of relatively high density. The basal portion of the ventricle protrudes between the suprachiasmatic nuclei. Between P1.5 mm and P2.1 mm the paraventricular nucleus occupies the dorsal half of the periventricular regions of the third ventricle. The cell types of the ependyma are similar, independent of whether they cover the periventricular or the paraventricular nuclei. The structure of the ventricular wall becomes more undulated.

The ependyma on the tuberal portion of the third ventricle is heterogeneous (Fig. 36). Topographic connections between the various types of ependyma and hypothalamic nuclei can be observed only in a few cases. The structure of the ependyma on the ventricular surface of the median eminence (on the floor of the third ventricle) has beenconsidered in Fig. 35. The ependyma overlying the arcuate nucleus is not constituted only by the so-called "arcuate tanycyte zone" but also by the monolayer and irregular tanycyte zones. Three zones of the tuberal ependyma (Fig. 35) are related to the ventromedial nucleus, while the more dorsal cuboidal ependyma is connected with the dorsomedial nucleus (Fig. 2B/19–23).

The ependyma in the "connecting part" is surrounded by the posterior periventricular area (Fig. 2B/24–26). The cell density is too low to form a real nucleus in that part of the hypothalamus. Dense neuropil, fibres, and various types of cells can be identified under the cuboidal ependyma layer.

The inframamillary recess penetrates deeply into the mamillary body. The roof of the recess reaches the dorsal premamillary nucleus and the medial mamillary nucleus (Fig. 2B/24–26). The tanycyte ependyma on the side wall of the recess is mainly localized to the territory of the arcuate nucleus. Basal processes of tanycytes pass through the aggregation of cells to the caudal portion of the median eminence.

CEREBRAL AQUEDUCT AND MESENCEPHALIC VENTRICLE

Topographic coordinates

The ependymal surface of the cerebral aqueduct and of the mesencephalic ventricle is relatively small, only 10% of the total ventricle surface (Table 1).

In Fig. 59 the schematic drawings of a series of frontal sections of the cerebral aqueduct (from section 26, P4.5 mm to section 37, P7.8 mm) and of the mesencephalic ventricle (from section 38 to section 42, P8.1 mm–P9.3 mm) is shown.

The sections are identical with those demonstrated in Fig. 2B (sections 26 to 42).

The rostral end of this part of the ventricle system is formed by the region of the subcommissural organ (SCO) that is localized on the ventral surface of the posterior commissure (CP).

The ependyma of the cerebral aqueduct can be divided into several parts. The rostral part is covered dorsally by cylindrical and cuboidal ependyma and ventrally by cuboidal ependyma. The central portion of the cerebral aqueduct (Fig. 59, sections 32–37, P6.3 mm–P7.8 mm) is covered by cuboidal ependyma on the dorsal wall and by cylindrical one on the ventral wall. As a consequence of the numerous recesses, the surface of the ventral ependyma is relatively large, nearly 40% of the total ependymal surface of the cerebral aqueduct (Table 4).

From the dorsal wall of the cerebral aqueduct the mesencephalic ventricle opens dorsally (Fig. 59, sections 38–42, P7.8 mm–P9.3 mm). The ependyma on the dorsal wall is mono- to multilayered. The cells are cuboidal or columnar. The ependyma on the ventral wall is formed by cuboidal cells. The dorsal wall of the more caudal portion of the cerebral aqueduct (Fig. 59, sections 41–42, P9.0 mm–P9.3 mm) is covered by irregular cuboidal ependyma, and the ventral wall of this portion of the aqueduct by cuboidal to columnar ependyma.

The topography of these regions is indicated in Fig. 60. The rostrocaudal extensions and the sizes of the surfaces of the individual ependymal regions are summarized in Table 4.

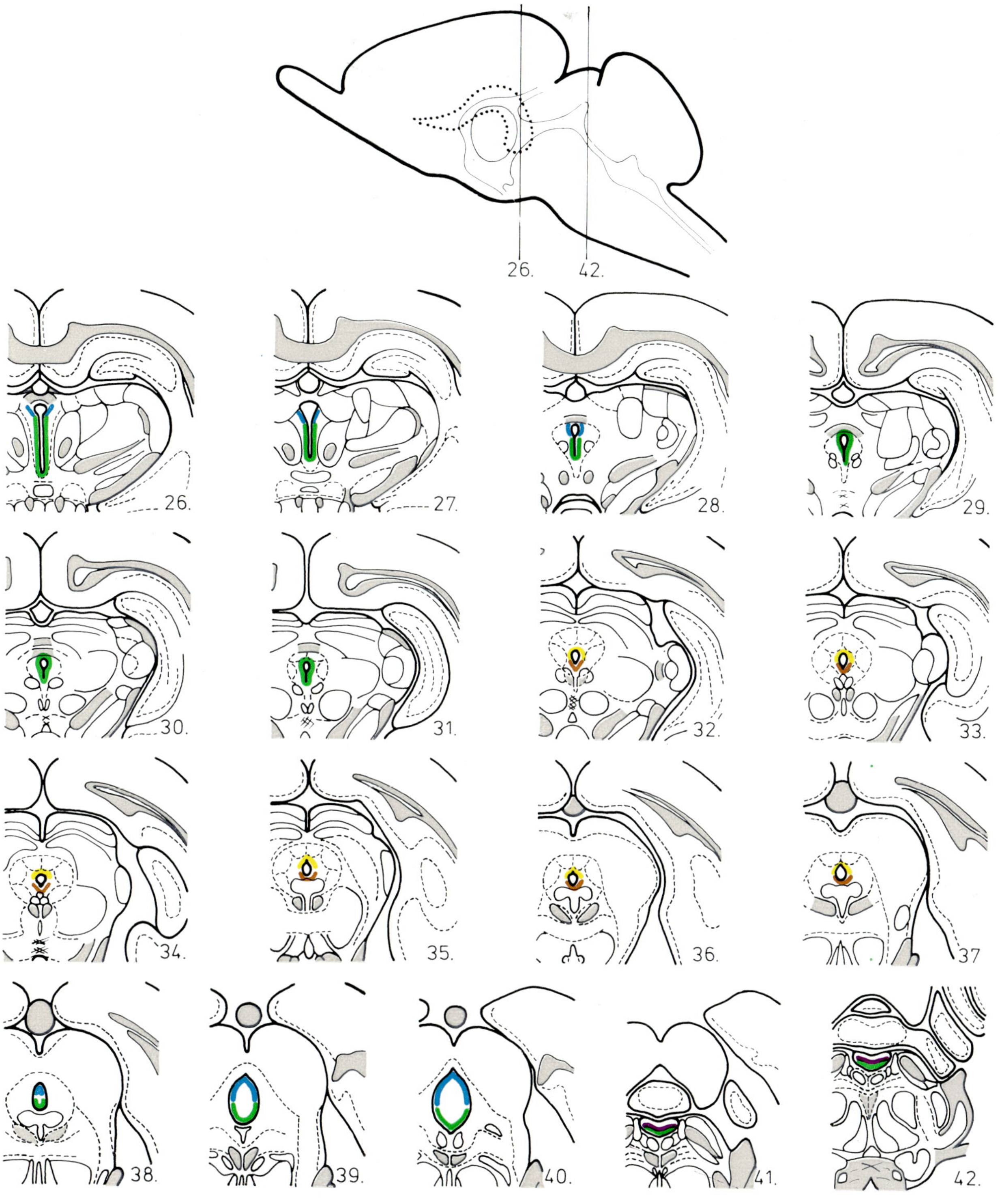

Fig. 59. Schematic drawing of frontal serial sections of the mesencephalic ependyma. Black indicates the SCO in sections 26 and 27 (P4.5 mm to P4.8 mm). The cylindrical ependyma on the dorsal wall of the aqueduct (sections 26–28; P4.5 mm to P5.1 mm) and of the mesencephalic ventricle (sections 38–42; P8.1 mm to P9.3 mm) is indicated by blue. The cuboidal ependyma of the ventral part of the aqueduct (sections 26–31, P4.5 mm to P6.0 mm and in sections 38–42, P8.1 mm to P9.3 mm) is indicated by green. The central portion of the cerebral aqueduct (sections 32–37, P6.3 mm to P7.8 mm) is covered by cuboidal ependyma on the dorsal wall (yellow) and by cylindrical one on the ventral wall (brown). For topography of those regions see Fig. 60. The dorsal wall of the caudal portion of the aqueduct (section 41–42; P9.0 mm to P9.3 mm) is covered by irregular cuboidal ependyma (violet) and the ventral wall by cuboidal ependyma (green)

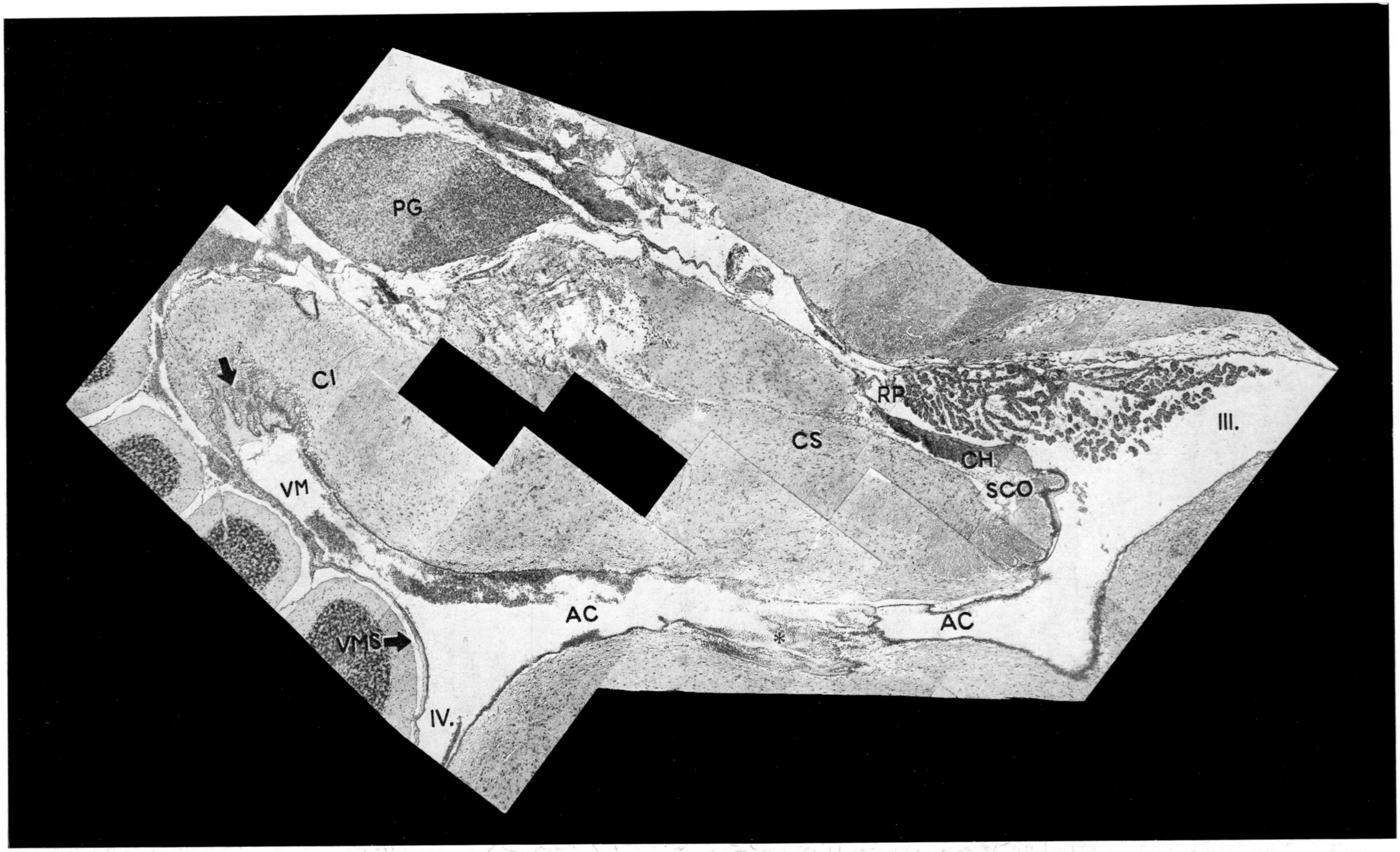

Fig. 60. Midsagittal section of the mesencephalic ventricle and the epithalamus including the pineal gland (PG). RP — pineal recess; III. — third ventricle; AC — cerebral aqueduct; VM — mesencephalic ventricle; IV. — fourth ventricle. The central part of the aqueduct, having a special type of ependyma, is indicated by an asterisk. The top of the mesencephalic ventricle is formed by large ependymal foldings (→). Magnification: ×36. For abbreviations see list

Structure of the ependymal areas of the cerebral aqueduct and mesencephalic ventricle

The subcommissural organ (SCO) is located on the dorsal, rostral, and ventral surfaces of the posterior commissure (Palkovits, 1965). According to its topographic relation to the posterior commissure (CP), the SCO can be divided into supracommissural, precommissural, subcommissural and retrocommissural parts. The cytology of the SCO is uniform in all parts, the subdivisions are only of topographic significance (Palkovits, 1965). The ependymal cells of the SCO are closely packed and have a large cytoplasm. The nuclei are localized in the basal part of the cells (Fig. 61A). The ependyma of the side walls partly covers the SCO cells (Fig. 61B). The product of the SCO cells condenses in the cerebrospinal fluid to form the liquor fibres (Reissner's fibre).

Caudally to the SCO the wall of the cerebral aqueduct assumes the shape of a fissure, and under the roof it forms small folds (Fig. 62A) as seen in frontal sections. The dorsal wall of the aqueduct is covered by columnar and cuboidal ependyma (Fig. 62B). The cell nuclei are round.

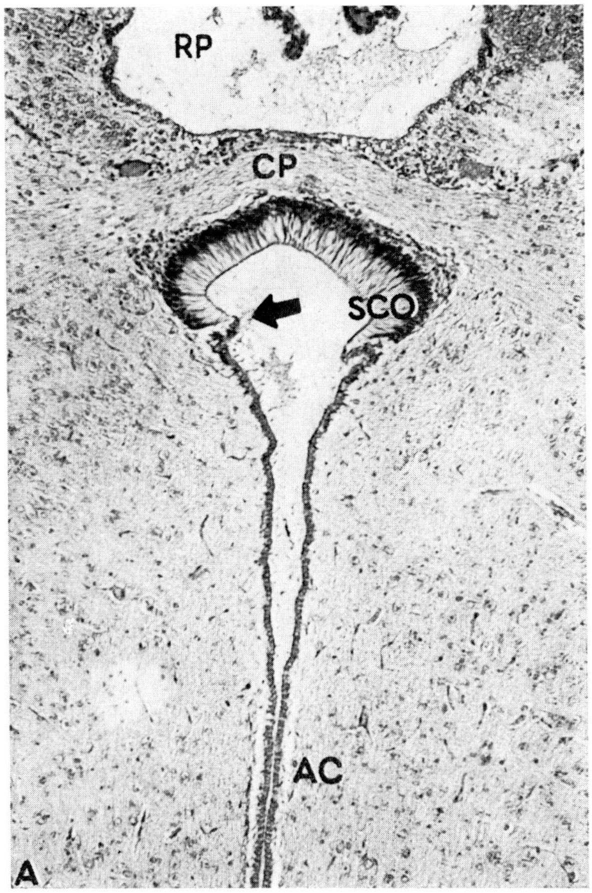

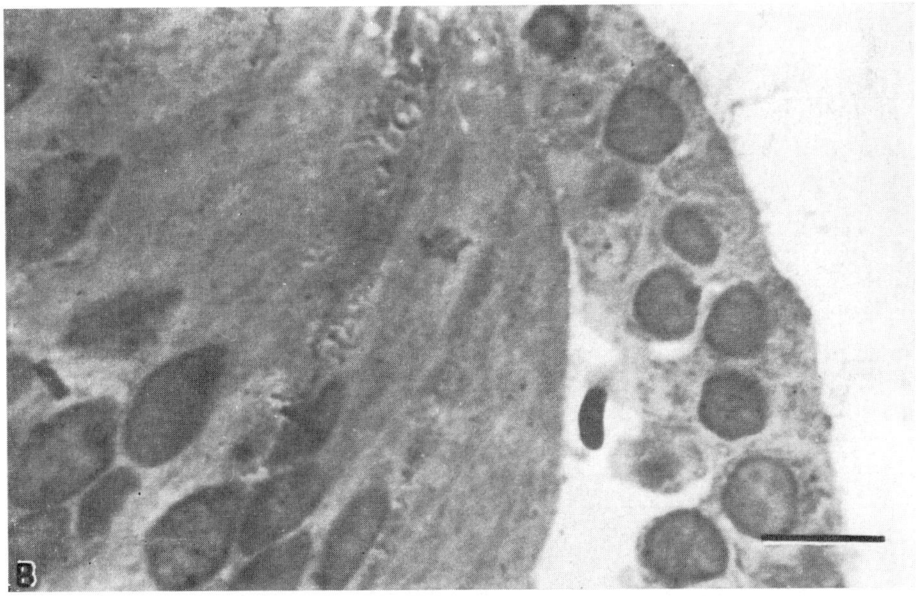

Fig. 61. **A** — the frontal section shows the rostral beginning of the cerebral aqueduct (AC); (section 26, P4.5 mm). CP — posterior commissure; SCO — subcommissural organ; RP — pineal recess (section 27, P4.8 mm). Magnification: ×60. **B** — the multilayered ependyma is attached to the SCO. In **A** this region is indicated by an arrow. Bar scale = 10 μm

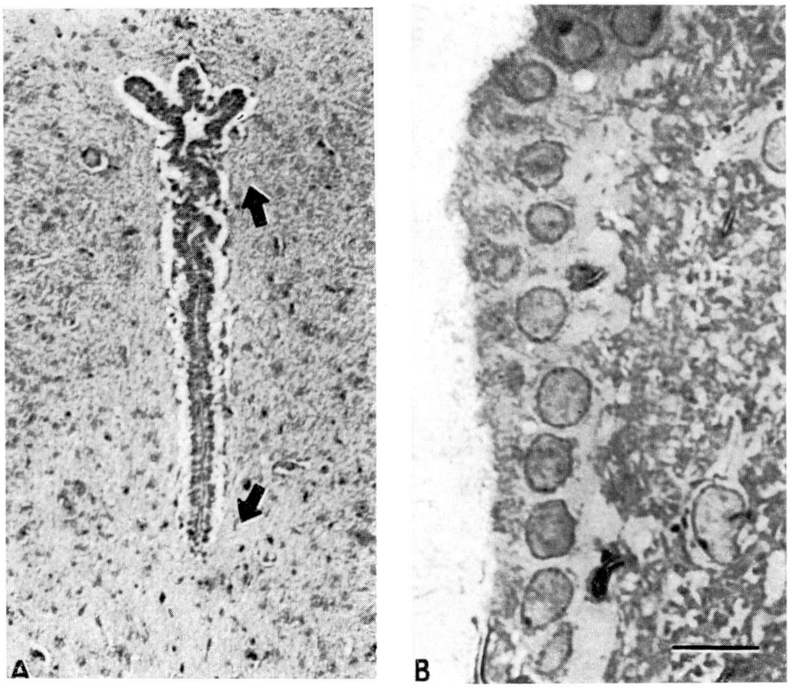

Fig. 62. Cuboidal ependyma from the ventral part of the cerebral aqueduct (→). **A** in Fig. 59 this region is green. **A** — magnification: ×100; **B** — bar scale = 10 µm

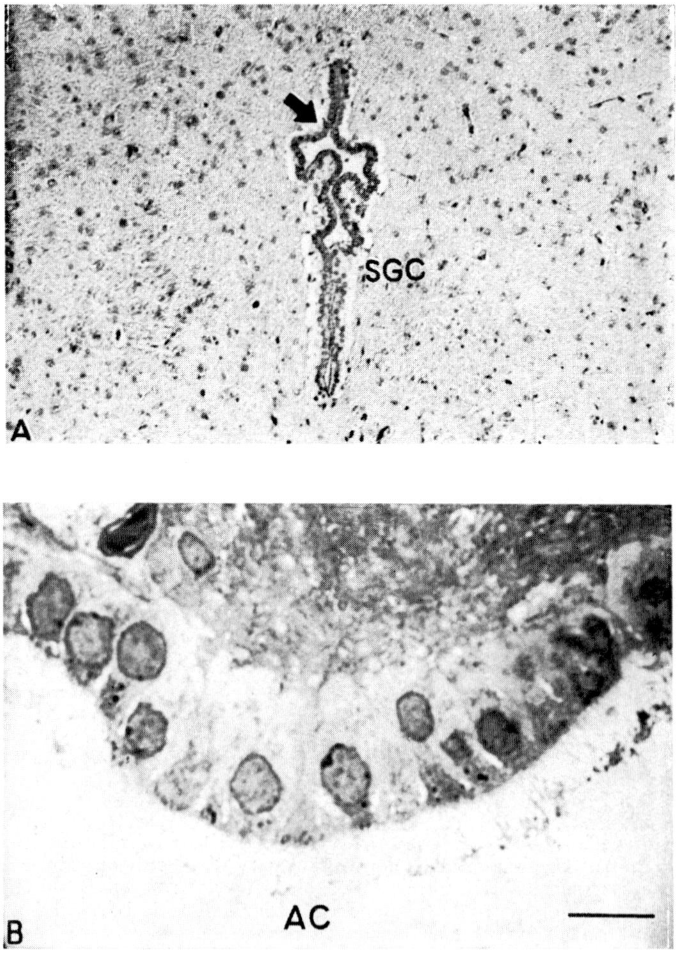

Fig. 63. **A** — the dorsal wall of the cerebral aqueduct (→) and central gray matter (SGC). Magnification: ×100. **B** — the dorsal wall of the cerebral aqueduct (AC) is covered by columnar ependyma. Bar scale = 10 µm

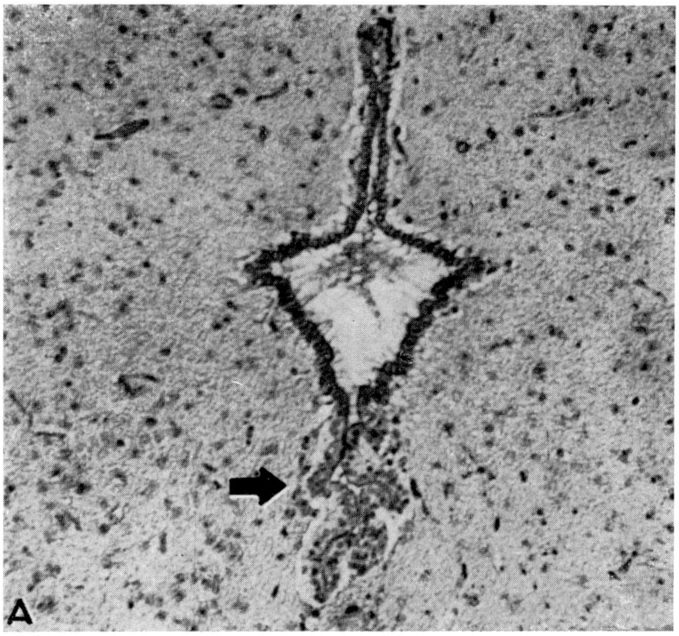

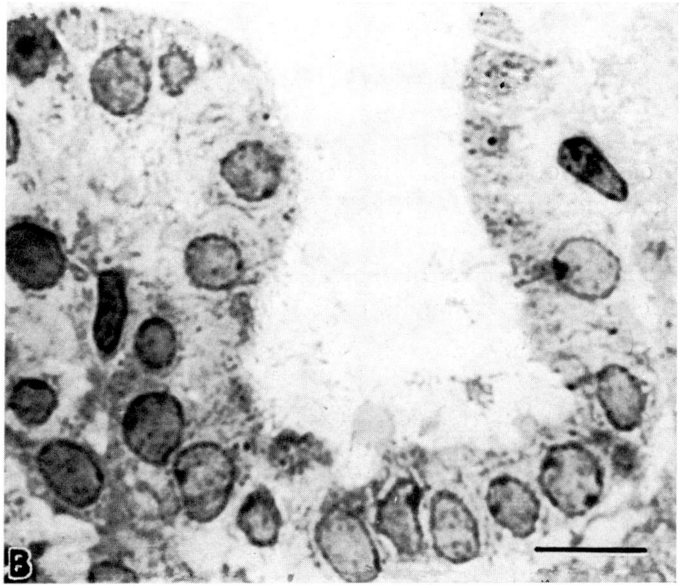

Fig. 64. Cuboidal to cylindrical ependyma from the ventral recess of the cerebral aqueduct (→). This region is indicated by an asterisk in Fig. 60. **A** — magnification: ×100; **B** — bar scale: 10 μm

The ventral wall of the aqueduct is formed by ciliated cuboidal cells with round nuclei (Fig. 63A, B).

The lumen of the central portion of the cerebral aqueduct forms several folds that protrude from the ventral wall (Fig. 64A). Cuboidal to columnar ependymal cells cover the walls of the ventral recesses of the aqueduct (Fig. 64B). The ependyma is ciliated. Bulbous protrusions are visible on the surface of the ependymal cells.

Caudalward, the lumen of the cerebral aqueduct becomes larger, the walls flatten, the lumen becomes oval and continues into the mesencephalic ventricle. The wall of this ventricle is generally formed by monolayered, occasionally by multilayered ependyma. On top, the ependyma is formed by columnar cells (Fig. 65A), and by columnar (Fig. 65B) and cuboidal cells on the side walls. The ependymal cells are covered by microvilli and cilia (Fig. 66A). The basal

Table 4

VENTRICLE SYSTEM OF THE LOWER BRAIN STEM

	Extension* Rostral (mm)	Extension* Caudal (mm)	Surface mm²	Surface fraction % 100% = 26.26 mm²
Cerebral aqueduct	P4.5	P9.3	*6.47*	*100.0*
dorsal cylindric ependyma	P4.5	P6.0	0.66	10.2
ventral cuboidal ependyma	P4.5	P6.0	2.52	38.9
dorsal wall in the central part	P6.3	P7.8	0.83	12.8
ventral wall in the central part	P6.3	P7.8	0.96	14.8
caudal ventral wall	P8.1	P9.3	1.50	23.2
Mesencephalic ventricle	P8.1	P9.3	*1.42*	
Fourth ventricle	P9.6	P13.8	*17.73*	*100.0*
ependyma on the roof	P9.6	P11.4	4.29	24.2
lamina epithelialis of the tela choroidea of the fourth ventricle	P11.4	P13.8	3.76	21.2
ventral wall			9.68	54.6
cuboidal ependyma	P9.6	P13.8	8.42	47.5
flattened ependyma	P9.6	P11.4	1.26	7.1
Central canal (medullary part)	P13.8	P15.9	*0.64*	

* 0 = bregma level, P = caudal to the bregma.

processes of the ependymal cells lie on the surrounding neuropil without any demarcation (Fig. 66B).

The total surface of the mesencephalic ventricle is relatively small, only 1.42 mm² in the rat (Table 4).

Cell groups in the vicinity of the cerebral aqueduct and mesencephalic ventricle

The rostral end of the ependyma of the cerebral aqueduct is related to the subcommissural organ (Fig. 2B/26–27). Along the course of the aqueduct the ependyma is surrounded by the central gray matter. This is a most heterogeneous gray matter incorporating circumscribed cell groups related to the ependyma. Large motoneurons of the oculomotor nerve (Fig. 2B/33–34) and cells of the dorsal raphe nucleus are subjacent the ventral wall of the aqueduct.

The mesencephalic ventricle and its ependymal surface invaginating into the inferior colliele are surrounded by a periventricular gray matter that does not differ from the mesencephalic central gray.

THE FOURTH VENTRICLE AND THE BEGINNING OF THE CENTRAL CANAL

Topographic coordinates

Caudal to the mesencephalic ventricle, the lumen of ventricle system continues first in the fourth ventricle, then in the central canal.

The surface of the fourth ventricle is relatively large, it amounts to one quarter of the total ventricular surface (Table 1). The surface of the medullary part of the central canal is small, less than 1 mm² in the rat. The fourth ventricle extends from section 43 (P9.6 mm) to section 57 (P13.8 mm). The area postrema is localized in sections 58 (P14.1 mm) and 59 (P 14.4 mm). The central canal begins from section 60 (P14.7 mm) that corresponds to the level of the obex. In Fig. 68 the frontal sections of the pontine and medullary ventricle system (section 42, P9.3 mm to section 64, P15.9 mm) are demonstrated.

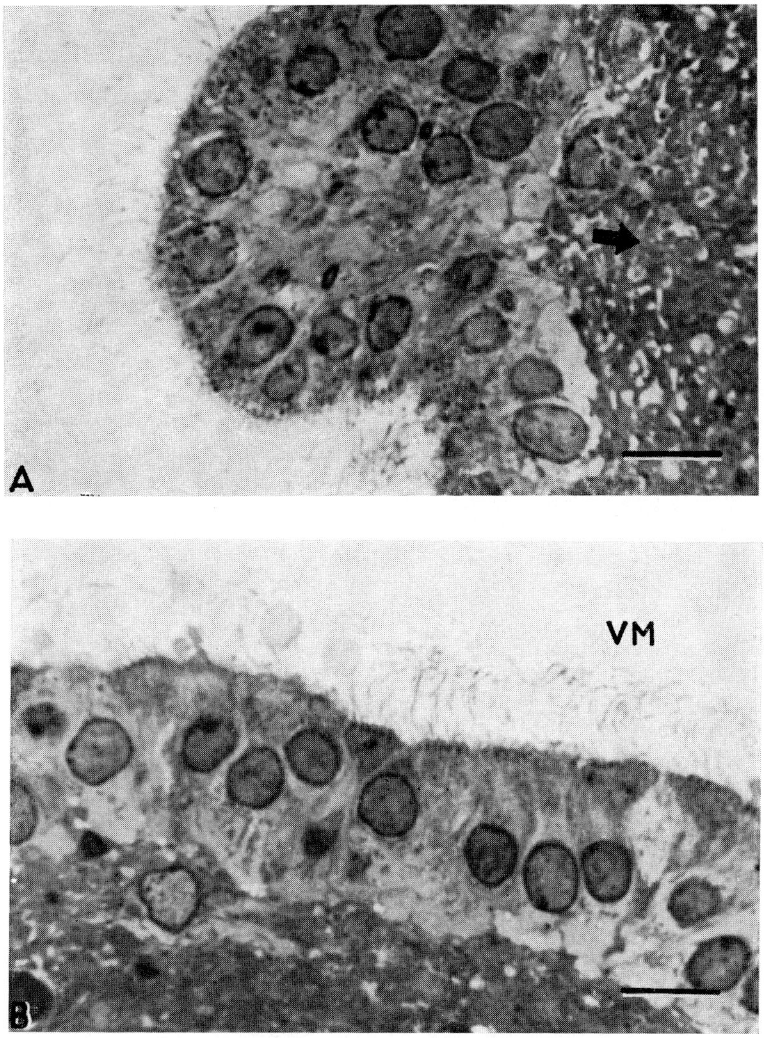

Fig. 65. Cylindrical ependyma from the top (**A**) and from the side wall (**B**) of the mesencephalic ventricle (VM). This region is indicated by blue in sections 38–42 in Fig. 59. Bar scale = 10 μm

The ependymal lining can be divided into the following regions: the irregular ependyma of the roof of the fourth ventricle, the flat ependyma of the lateral portion of the ventral wall, the cuboidal ependyma of the central portion of the roof, the cuboidal ependyma of the ventral wall of the fourth ventricle, and the ependyma of the beginning of the central canal.

The surface of the ventral wall is nearly as large (9.68 mm^2) as the surfaces of the roof ependyma and the epithelial lamina of the tela choroidea of the fourth ventricle together (Table 4). (Total surface of the epithelial lamina could not be measured on the convoluted surface of the plexus. Data are obtained only from the basal, relatively flat regions of the lamina.)

The topography of these regions is indicated in Fig. 67, the rostrocaudal extensions, and the sizes of their ventricular surfaces in Table 4.

Structure of the ependymal areas of the fourth ventricle and the beginning of the central canal

The ependymal cells lining the roof of the fourth ventricle show considerable variations, but on the whole the principal types of the ependymal cells are flattened and cuboidal. Flattened cells (Fig. 69A, B) show short, sparse cilia on their surface and elongated nuclei. Cuboidal cells (Fig. 70A) have longer cilia and larger, round nuclei. On the apical surface of the cells intraventricular nerve fibres can be observed (Fig. 70B).

The ependymal layer continues from the lateral parts of the roofs as lamina epithelialis onto the ventricle surface of the tela choroidea of the fourth ventricle (Fig. 71A, B).

The rostral half of the floor of the fourth ven-

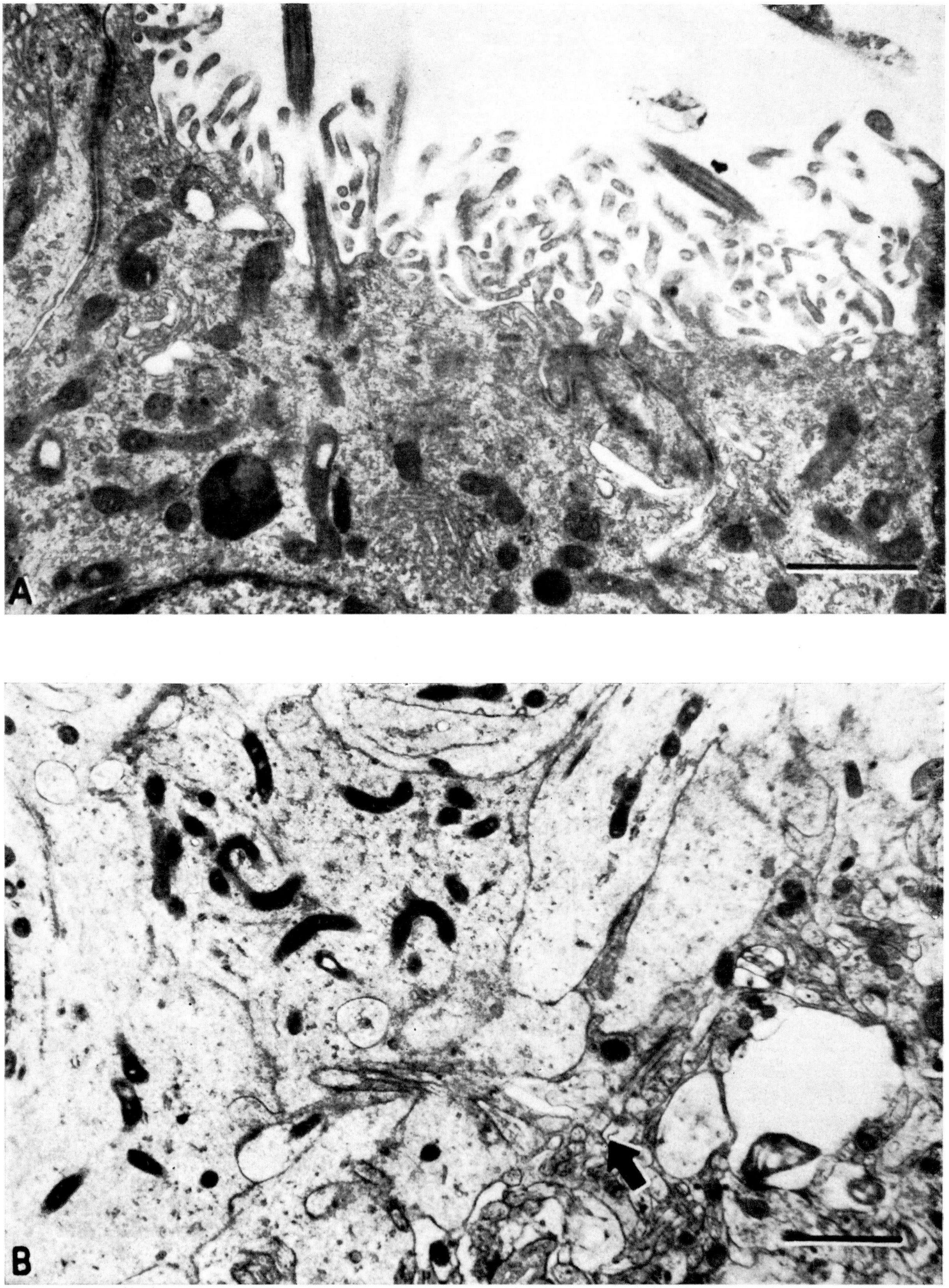

Fig. 66. Apical (**A**) and basal (**B**) parts of the ependymal cells of the mesencephalic ventricle. Microvilli and cilia are visible on the apical surface of the cells. Basal processes of the ependymal cell lie on the subependymal neuropil (→) without any demarcation. Bar scales: **A** = 2 μm; **B** = 1 μm

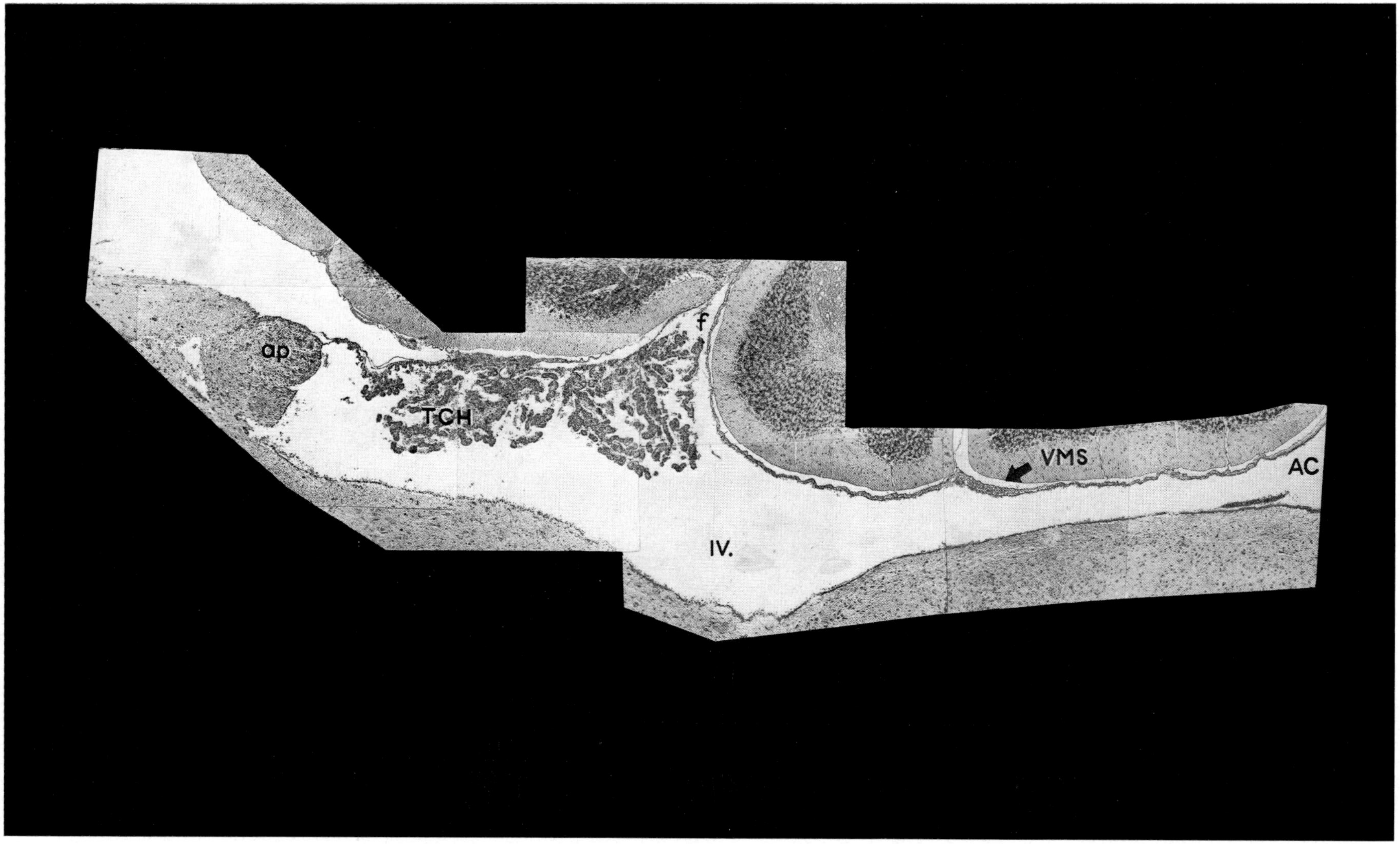

Fig. 67. Midsagittal section of the fourth ventricle (IV.) TCH — tela choroidea. For abbreviations see list

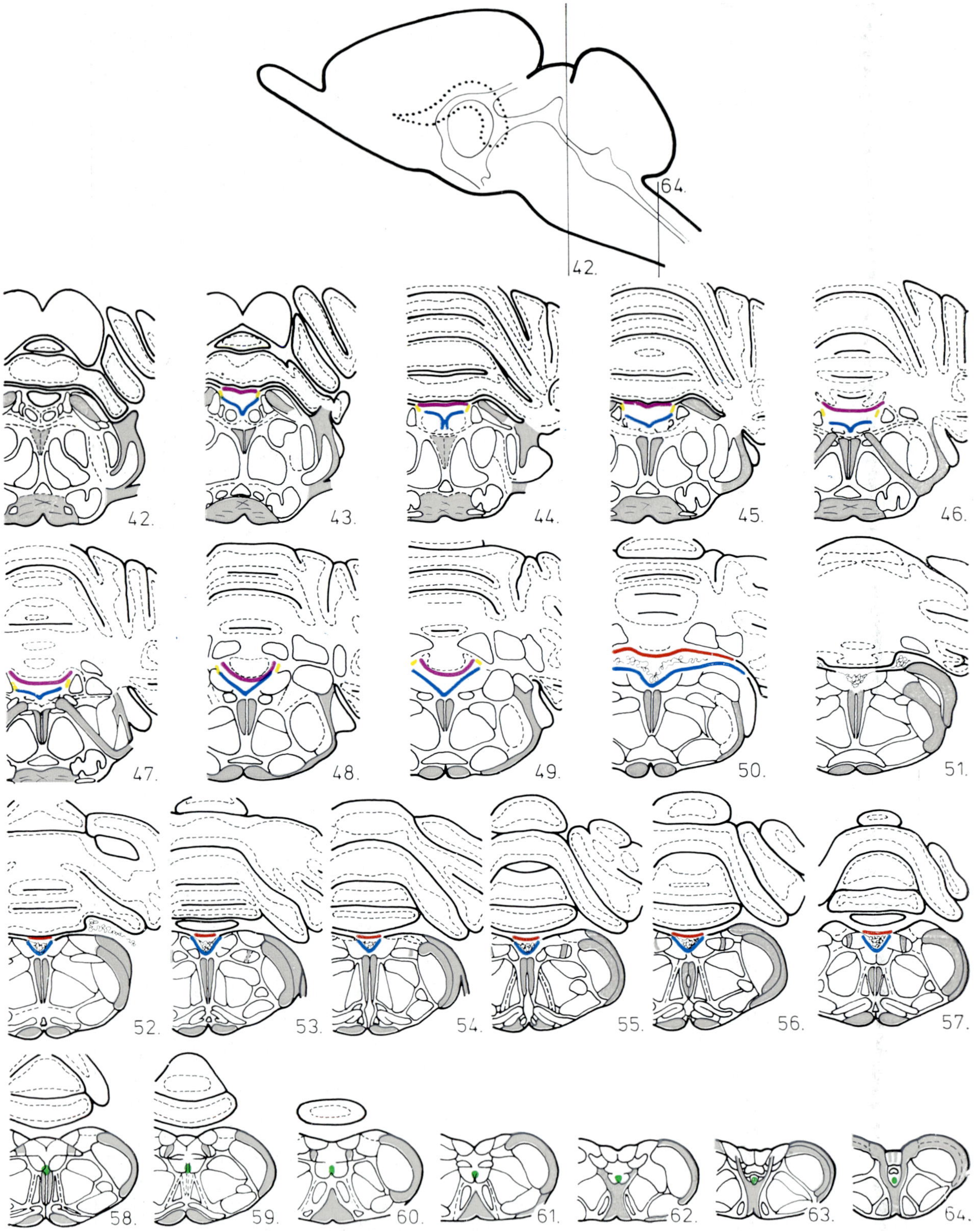

Fig. 68. Schematic drawing of the frontal sections of the pontine and medullary ventricle system. Violet – irregular, flat to cuboidal ependyma on the roof of the fourth ventricle (on the ventrical surface of the superior medullary velum), sections 43–49 (P9.6 mm to P11.4 mm). Red – epithelial lamina of the tela choroidea of the fourth ventricle (sections 50–57; P11.7 mm to P13.8 mm). In its rostral half, the lateral portion of the ventral wall is covered by flat ependyma (yellow in section 43–49; P9.6 mm to P11.4 mm). The central portion of the ventricle floor as well as the whole ventral wall of the caudal half of the fourth ventricle are composed of cuboidal ependyma (blue in section 43–57; P9.6 mm to P13.8 mm). Green – ependyma of the central canal (sections 58–64; P14.1 mm to P15.9 mm)

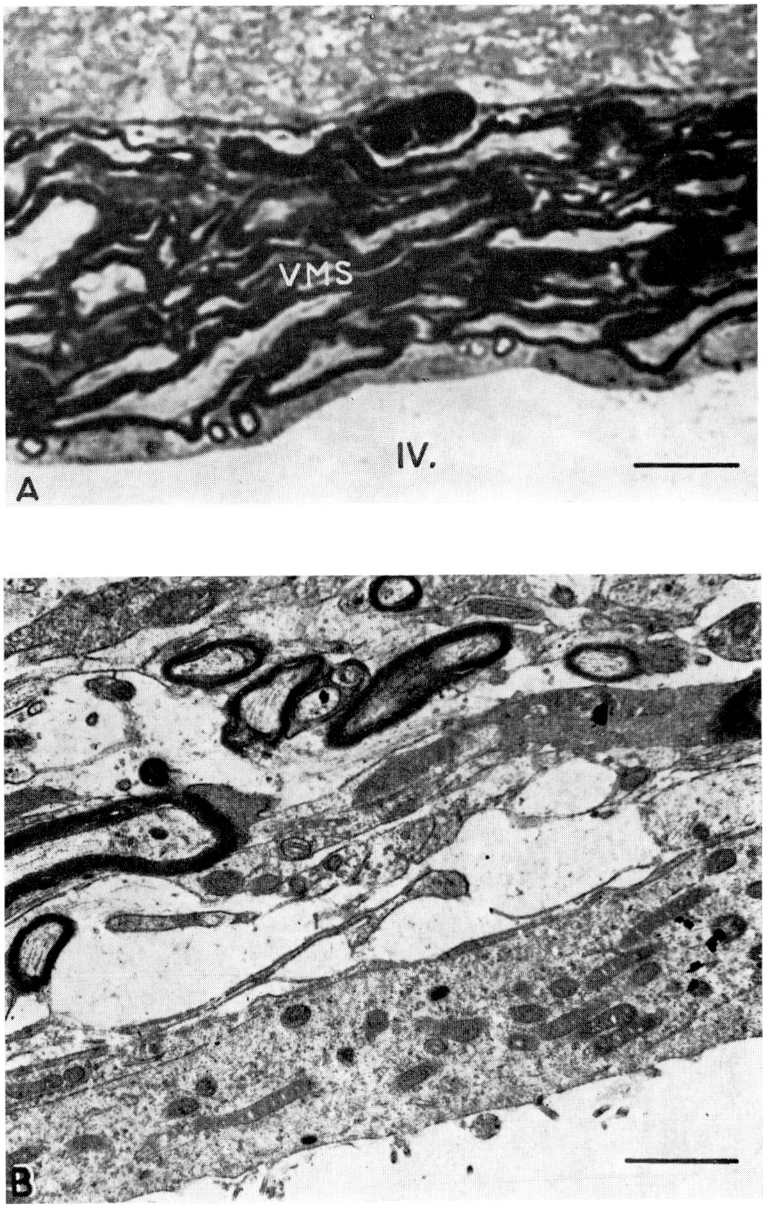

Fig. 69. Irregular flat ependyma from the roof of the fourth ventricle. VMS — superior medullary velum; IV. — fourth ventricle. Bar scales: **A** = 10 μm; **B** = 1 μm

tricle and the lateral portion of the ventral wall (Fig. 72) are formed by flat ependymal cells. The central portion of the ventricle floor and the ventral wall of the caudal half of the fourth ventricle are lined with cuboidal ependyma (Fig. 73).

The lateral recesses are lined with flattened and cuboidal ependymal cells.

Dorsal to the area postrema, cuboidal ependymal cells can be demonstrated.

The lumen of the fourth ventricle continues caudally in the central canal. We followed only the initial medullary part of the canal. It is lined with cuboidal or columnar ependymal cells with ovoid cell nuclei (Fig. 74).

In several places the ependymal lining forms folds extending to various distances in the lumen. The surface of the cells is covered by cilia. Among or upon the cilia a few spherical structures can be seen. In the surrounding tissue the subependymal cells occur in close contact with the adjacent ependymal cells.

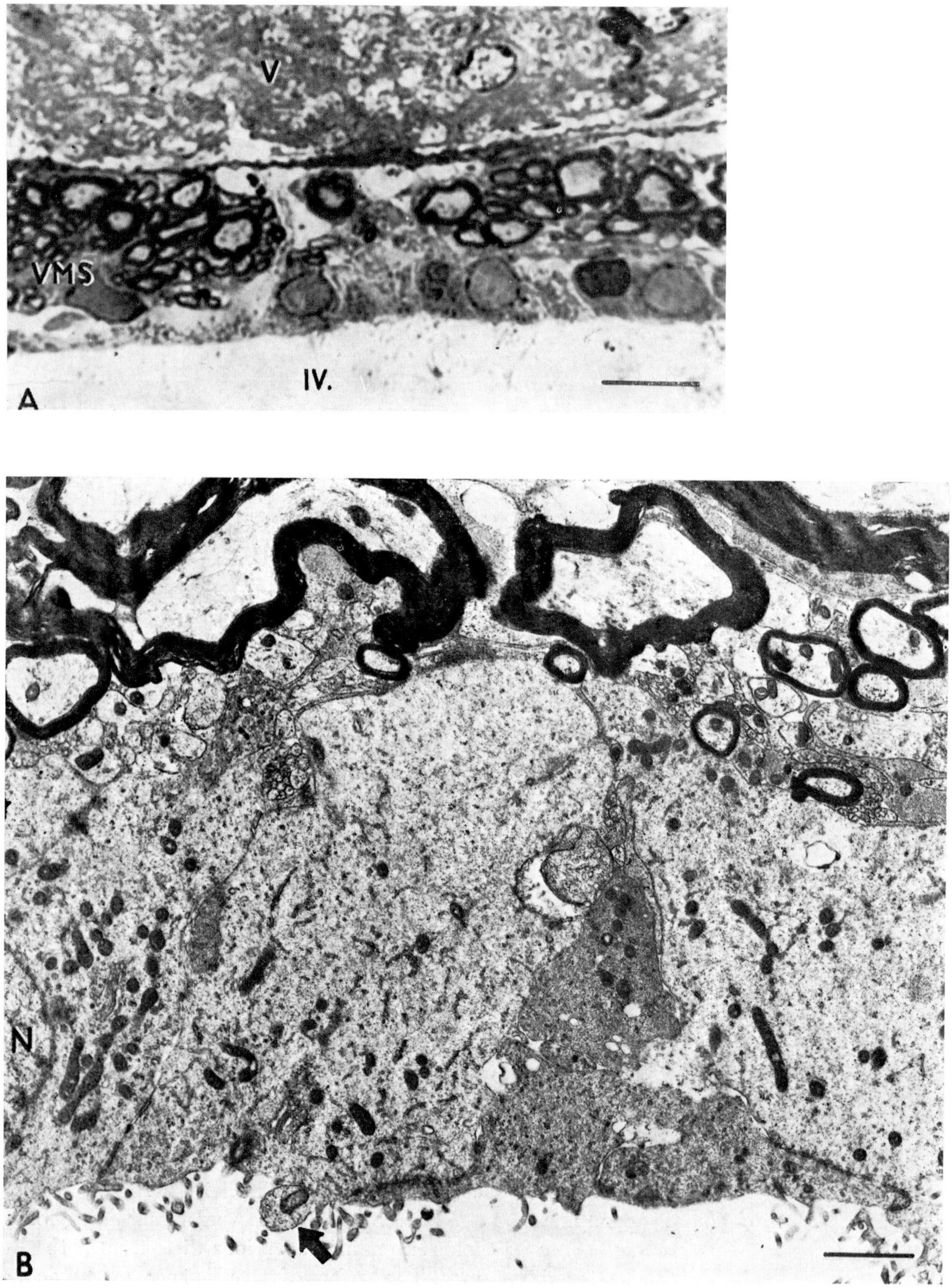

Fig. 70. **A** and **B** — Irregular cuboidal ependyma from the roof of the fourth ventricle. VMS — superior medullary velum; V — vermis cerebelli; IV. — fourth ventricle; N — cell nucleus. The intraventricular nerve fibre is indicated by an arrow. Bar scales: **A** = 10 μm; **B** = 1 μm

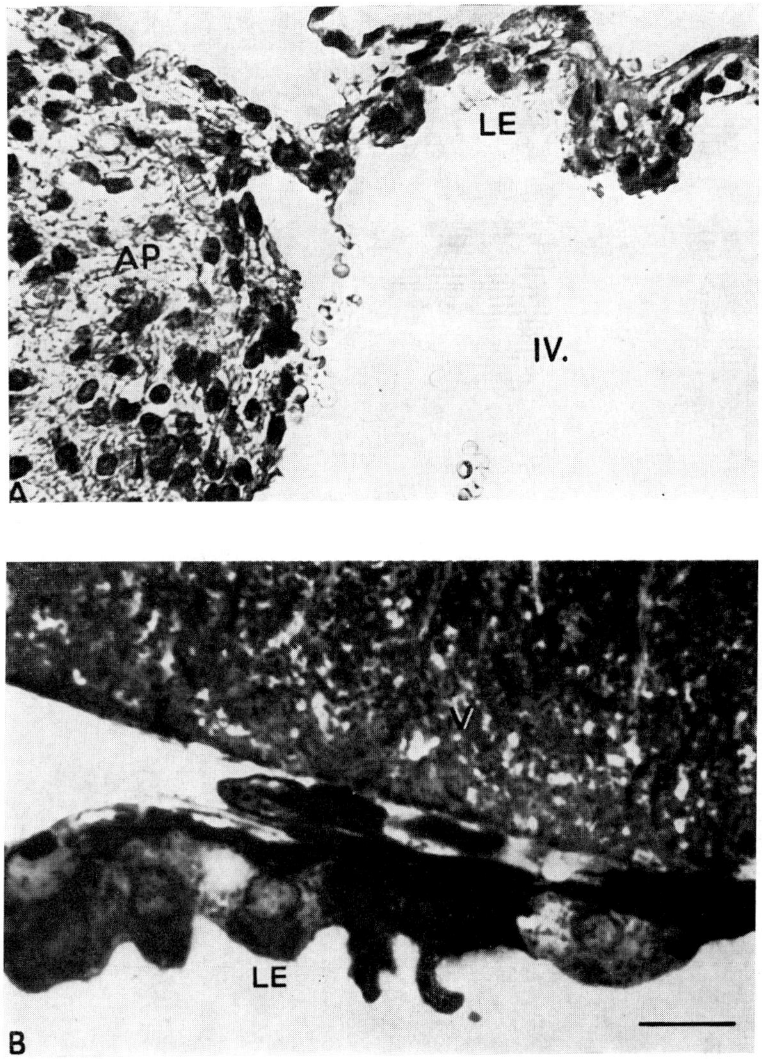

Fig. 71. **A** and **B** — lamina epithelialis (LE) on the ventricle surface of the tela choroidea of the fourth ventricle (IV.). **A** — magnification: ×400; **B** — bar scale; 10 μm. For abbreviations see list

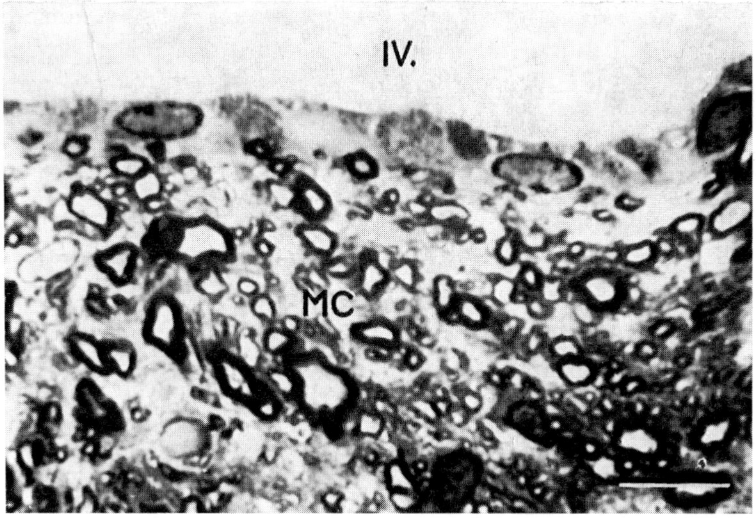

Fig. 72. Flat ependyma from the ventrolateral portion of the fourth ventricle (IV.). Topography of this region is indicated by yellow in Fig. 68. MC — cross sections of myelinated fibre from the cerebellar medulla. Bar scale = 10 μm

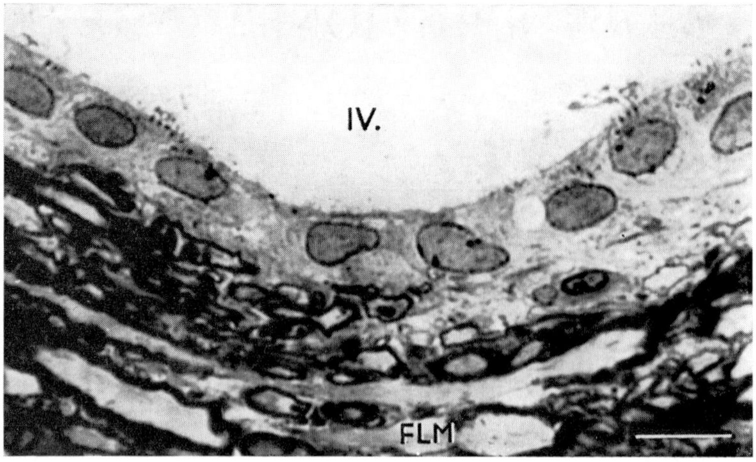

Fig. 73. Cuboidal ependyma from the floor of the fourth ventricle (IV.). FLM — cross sections of fibres of the medial longitudinal fascicle. Bar scale = 10 μm

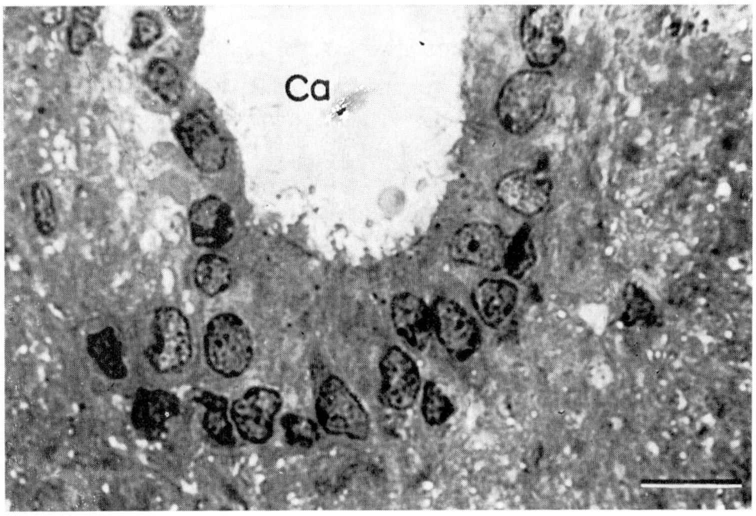

Fig. 74. Central canal (Ca) surrounded by irregular mono- to multilayered cuboidal ependyma. Bar scale = 10 μm

Cell groups in the vicinity of the fourth ventricle and at the beginning of the central canal

Practically no periventricular structure can be found under the ependyma of the roof of the ventricle. It covers the superior medullary velum along its full length (Fig. 2B/41–47).

It is noteworthy that the ependyma of the lateral edge of the ventricle differs from the other types and has a topographic connection with the locus coeruleus (Fig. 2B/41–45).

The ventral wall of the fourth ventricle covers the pontine tegmentum (Fig. 2B/41–55). This gray substance contains various types of cells and cell groups (dorsal tegmental nucleus, salivatory nucleus, vestibular nuclei, prepositus hypoglossal nucleus, a part of the nucleus tractus solitarii). Neither specific structural nor topographic connections exist between the ventral wall's ependyma and the above-mentioned cell groups.

The central canal starts from the fourth ventricle at the obex. The medullary portion of the canal is surrounded by various cell groups. In the rostrocaudal direction they are found in the following order: area postrema, nucleus tractus solitarii, dorsal vagal nucleus, intercalate nucleus, motor nucleus of the hypoglossal nerve (Fig. 2B/58–64).

GENERAL CONSIDERATIONS

In all vertebrates the walls of the brain ventricles and of the central canal are covered by a special ependymal lining. The ependyma consists of three major components: a layer of (1) ependymal cells, (2) subependymal and (3) supraependymal structures. There are regions where one or two of the above layers are missing. The ependymal cell layer itself is always present. Sometimes the term of ependyma is used to designate the ependymal cells or the layers of these cells. However, the name ependym or ependyma can be used for all components (ependymal, supraependymal and subependymal) taking part in the construction of the ventricle walls.

The ependymal cells, as specialized glial cells, represent the basic component of the ependymal layer. They differ in shape and size according to their topography. Flat, cuboidal and columnar ependymal cells can be distinguished. They may occur in a single layer or may form two or several layers. Their arrangement may be regular or irregular.

In combination with the ependymal cells, other glial elements, intraependymal capillaries, nerve fibres are found within the ependymal layer. There are regions in the ventricular walls where tanycytes can be observed. Thus, the ependymal layer exhibits great variations.

The subependymal layer is formed by glial elements of various thickness. The word "subependymal" has rather a topographic meaning than discrimination of cells according to their characteristic features. The term "subependymal cell" should be reserved for glial cells (it is unnecessary to use the name ependymo-glial cells) not including neurons even if they are located in the immediate vicinity of the ependymal layer.

The third component of the ependyma is constituted by supraependymal structures like cells, cell protrusions, nerve fibres and endings.

The ependyma is frequently described to be a complicated mosaic. From the morphological point of view, the real meaning of "mosaic" concerns the complexity of the ventricular surface of the ependyma. The occurrence of microvilli, cilia, globular or pseudopodia-like protrusions yields indeed a colourful mosaic structure on the ventricular surface but does not automatically refer to parallel variations in the depth of the ependyma. In addition, there are great differences inside the wall due to the presence or absence, and to the varying relative proportion of the ependymal and subependymal elements. Therefore, as the ventricular ependyma is rather heterogeneous, regional characteristics can be demonstrated. There are ependymal structures like the choroid plexus epithelium, ependymal cells on the ventricular surface of the circumventricular organs, etc., that markedly differ from one another.

The physiological role of the ependyma in the central nervous system is still unknown. The diversity of the morphological appearance of the ependyma does not necessarily mean a functional heterogeneity of the same degree. The presence of tanycyte ependyma in the third ventricle suggests the possible role of the ependyma in transport mechanisms between the ventricle and the neighbouring capillaries. Basal processes of tanycytes interconnect the internal and external CSF spaces. As to the tanycytes, two topographic observations have to be emphasized. (1) The topographic occurrence of the tanycytes in the hypothalamic ventricular wall corresponds to the so-called hypophysiotrophic area (Mészáros et al., 1969). This area, in contrast to former explanations, seems to be a trophic area, having a special transport capacity and connection between the endocrine hypothalamus and the pituitary gland (Ambach et al., 1977; Palkovits, 1978; Makara et al., 1980).

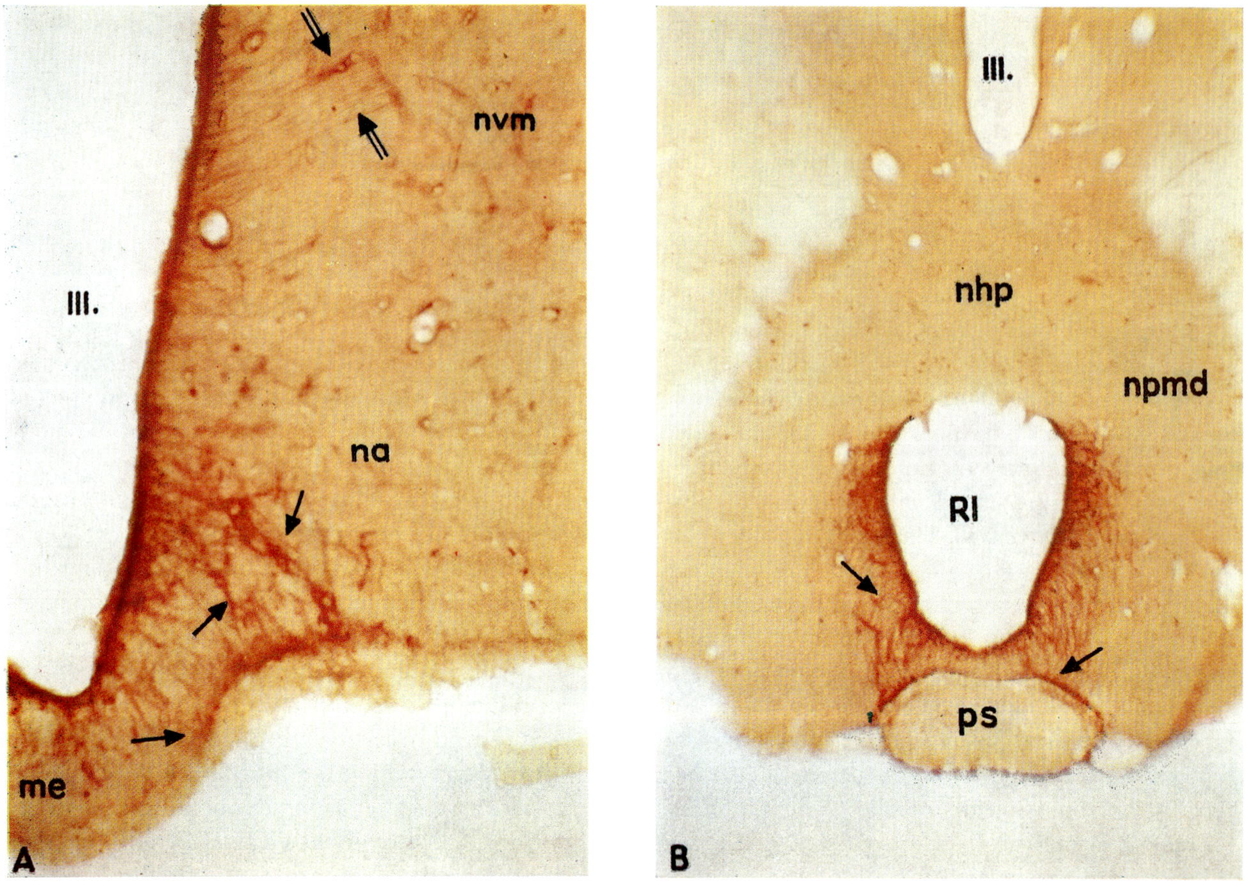

Fig. 75. Tanycyte processes in the middle portion of the third ventricle (**A**) and at the inframamillary recess (**B**). Frontal sections, nonspecific cholinesterase staining. Arrows indicate tanycytes between the ventricles and the basal surface of the diencephalon. Double arrows indicate tanycytes between the third ventricle and the hypothalamic capillaries. Magnification: **A** = ×50; **B** = ×30. For abbreviations see list

The possible role of tanycytes in these mechanisms cannot be excluded. It is worthwhile to mention that this particular region lies outside the blood-brain barrier. (2) In contrast to many recent descriptions, we find that tanycyte processes terminate not only in the pericapillary spaces of the median eminence-pituitary stalk but also in that of other hypothalamic capillaries, mostly in the regions of the arcuate and ventromedial nuclei, and of the retrochiasmatic area (Fig. 75). It is known that capillaries from these regions do not reach the pituitary gland but are collected by the basal veins draining into the systemic circulation instead of the pituitary portal vessels (Ambach et al., 1976; Ambach and Palkovits, 1977). Thus, to the tanycytes a role may be attributable in establishing a communication between the third ventricle and both the median eminence-pituitary circulation and the systemic circulation.

The third possible transport of material exchange between the CSF and the subependymal brain structures should occur through the ependyma itself. Substances injected intraventricularly can reach the deeper areas passing through the ependymal wall. Active cellular transport cannot be excluded. The different concentrations in the periventricular regions of intraventricularly injected substances (Verhoef et al., 1977) indicate the active participation of the related ependyma in the transport processes.

The recently developed preparation technique to isolate and purify ependymal cells (Manthorpe et al., 1977) facilitates the characterization of these cells. This preparation should be useful in determining the biochemical nature of the ependyma in various regions of the brain under normal and experimental conditions. Further biochemical and physiological experiments should be undertaken to elucidate the physiological function of this specific element of the central nervous system.

REFERENCES

AGDUHR, E. (1932): Choroid plexus and ependyma. In: W. Penfield (ed.). *Cytology and Cellular Pathology of the Nervous System.* Vol. II. Hoebner, New York pp. 537–573.

AGHAJANIAN, G. K., GALLAGER, D. W. (1975): Raphe origin of serotoninergic nerves terminating in the cerebral ventricles. *Brain Res.* **88**, 221–231.

AKERT, K., PFENNIGER, K., SANDORI, C. (1967): The fine structure of synapses in the subfornical organ of the cat. *Z. Zellforsch.* **81**, 537–556.

AKMAYEV, I. G., FIDELINA, O. V. (1974): Morphological aspects of the hypothalamic-hypophyseal system. V. The tanycytes: their relation to the hypophyseal adrenocorticotrophic function. An enzymehistochemical study. *Cell Tissue Res.* **152**, 403–410.

AKMAYEV, I. G., FIDELINA, 10. V., KABOLOVA, Z. A., POPOV, A. P., SCHITKOVA, T. A. (1973): Morphological aspects of the hypothalamic-hypophyseal system. IV. Medial basal hypothalamus. An experimental morphological study. *Z. Zellforsch.* **137**, 493–512.

ALLEN, D. J., LOW, F. N. (1973): The ependymal surface of the lateral ventricle of the dog as revealed by scanning electron microscopy. *Amer. J. Anat.* **137**, 483–489.

AMBACH, G., PALKOVITS, M. (1977): Die Blutversorgung des medialen basalen Hypothalamus. Anwendung einer Injektionsmethode, die eine direkte Unterscheidung von Arterien und Venen ermöglicht. *Verh. Anat. Ges.* **71**, 1455–1462.

AMBACH, G., PALKOVITS, M., SZENTÁGOTHAI, J. (1976): Blood supply of the rat hypothalamus. IV. Retrochiasmatic area, median eminence, arcuate nucleus. *Acta Morphol. Acad. Sci. Hung.* **24**, 93–118.

ARNOLD, W. (1970): Über eigentümliche neuronale Zellenelemente im Ependym des zentral Kanals von *Salamandra maculosa. Z. Zellforsch.* **105**, 176–187.

BAKER, B. L., DERMODY, W. C., REEL, J. R. (1975): Distribution of gonadotropin-releasing hormone in the rat brain as observed with immunocytochemistry. *Endocrinology* **97**, 125–135.

BARRY, J. (1977): Immunofluorescence study of LRF neurons in man. *Cell Tissue Res.* **181**, 1–14.

BARRY, J., CARETTE, B. (1975): Immunofluorescence study of LRF neurons in primates. *Cell Tissue Res.* **164**, 163–178.

BARRY, J., DUBOIS, M. P. (1976): Immunoreactive neurosecretory pathways in mammals. *Acta Anat. (Basel)* **94**, 497–503.

BARRY, J., DUBOIS, M. P., POULAIN, P. (1973): LRF-producing cells of the mammalian hypothalamus. *Z. Zellforsch.* **146**, 351–366.

BARTONÍČEK, V., LOJDA, Z. (1964): Topochemistry of enzymes of choroid plexus and ependyma of four animal species. I. Hydrolytic enzymes. *Acta Histochem. (Jena)* **19**, 357–368.

BARTONÍČEK, V. LOJDA, Z. (1966): Topochemistry of enzymes of choroid plexus and ependyma of four animal species. II. Diaphorases and dehydrogenases. *Acta Histochem. (Jena)* **23**, 118–126.

BEN-JONATHAN, N., MICAL, R. S., PORTER, J. C. (1974): Transport of LRF from CSF to hypophysial portal and systemic blood and the release of LH. *Endocrinology* **95**, 18–25.

BLAKEMORE, W. F. (1969): The ultrastructure of the subependymal plate in the rat. *J. Anat.* **104**, 423–433.

BLAKEMORE, W. F., JOLLY, R. D. (1972): The subependymal plate and associated ependyma in the dog. An ultrastructural study. *J. Neurocytol.* **1**, 69–84.

BLEIER, R. (1975): Surface fine structure of supraependymal elements and ependymas of hypothalamic third ventricle of mouse. *J. Comp. Neurol.* **161**, 555–568.

BLINZINGER, K. (1962): Elektronmikroskopische Untersuchungen am Ependym der Hirnventrikel des Goldhamsters (*Mesocricetus aureatus*). *Acta Neuropathol. (Berl.)* **1**, 527–532.

BÖHME, G., BODO, F. (1967): Die Binnenräume des Gehirns von Ratte und Maus. *Acta Anat. (Basel)* **68**, 199–206.

BOOZ, K. H. (1975): Secretory phenomena at the ependyma of the IIIrd ventricle of the embryonic rat. *Anat. Embryol. (Berl.)* **147**, 143–159.

BOOZ, K. H., DESAGA, U., FELSING, T., FRANZ, H., STARK, M. (1972): Sub- und interependymale Basalmembranlabyrinthe am Zentralkanal der weissen Ratte. *Z. Zellforsch.* **132**, 217–229.

BOOZ, K. H., DESAGA, U., FELSING, T. (1974): Über die Einstehung der Basalmembranlabyrinthe. Eine licht- und elektronen-mikroskopische Untersuchung. *Z. Anat. Entwickl.-Gesch.* **143**, 185–203.

BOULDER COMMITTEE (1970): Embryonic vertebrate central nervous system: revised terminology. *Anat. Rec.* **166**, 257–262.

BRAWER, J. R. (1972): The fine structure of the ependymal tanycytes at the level of the arcuate nucleus. *J. Comp. Neurol.* **145**, 25–42.

Brawer, J. R., Lin, P. S., Sonnenschein, C. (1974): Morphological plasticity in the wall of the third ventricle during the estrous cycle in the rat. A scanning electron microscopic study. *Anat. Rec.* **179,** 481–490.

Brightman, M. W. (1961): The fine structure of ciliated ependyma. *Anat. Rec.* **139,** 210–211.

Brightman, M. W., Palay, S. L. (1963): The fine structure of ependyma in the brain of rat. *J. Cell Biol.* **19,** 415–439.

Brownfield, M. S. (1977): Involvement of the magnocellular neurosecretory system in the mediation of diverse brain and endocrine functions. *Anat. Rec.* **187,** 541.

Brownstein, M., Palkovits, M., Kizer, J. S. (1975): On the origin of luteinizing hormone-releasing hormone (LH-RH) in the supraoptic crest. *Life Sci.* **17,** 679–682.

Brownstein, M. J., Palkovits, M., Saavedra, J. M., Kizer, J. S. (1975): Tryptophan hydroxylase in the rat brain. *Brain Res.* **97,** 163–166.

Bruni, J. E., Montemurro, D. G., Clattenburg, R. E., Singh, R. P. (1972): A scanning electron microscopic study of the ependymal surface of the third ventricle of the rabbit, rat, mouse and human brain. *Anat. Rec.* **174,** 407–420.

Bruni, J. E., Montemurro, D. G., Clattenburg, R. E. (1977): Morphology of the ependymal lining of the rabbit third ventricle following intraventricular administration of synthetic luteinizing hormone-releasing hormone (LH-RH). A scanning electron microscopic investigation. *Amer. J. Anat.* **150,** 411–426.

Cancilla, P. A., Zimmerman, H. M., Becker, N. H. (1966): A histochemical and fine structure study of the developing rat choroid plexus. *Acta Neuropathol. (Berl.)* **6,** 188–200.

Chamberlain, J. G. (1974): Scanning electron microscopy of epiplexus cells (macrophages) in the fetal rat brain. *Amer. J. Anat.* **139,** 443–446.

Clementi, F., Marini, D. (1972): The surface fine structure of the walls of cerebral ventricles and of choroid plexus in cat. *Z. Zellforsch.* **123,** 82–95.

Coates, P. W. (1973): Supraependymal cells: light- and transmission electron microscopy extends scanning electron microscopic demonstration. *Brain Res.* **57,** 502–507.

Coates, P. W. (1977): The third ventricle of monkeys. Scanning electron microscopy of surface features in mature males and females. *Cell Tissue Res.* **177,** 307–316.

Colmant, H. J. (1967): Über die Wandstruktur des dritten Ventrikels der Albinoratte. *Histochemie* **11,** 40–61.

Dalen, H., Schlapfer, W. T., Mamoon, A. (1971): Cilia on cultured ependymal cells examined by scanning electron microscopy. *Exp. Cell Res.* **67,** 375–379.

Dermietzel, R., Meller, K., Tetzlaff, W., Waelsch, M. (1977): *In vivo* and *in vitro* formation of the junctional complex in choroid epithelium. *Cell Tissue Res.* **181,** 427–441.

Desaga, U. (1971): Licht- und elektronmikroskopische Untersuchungen über die Reversibilität von Kernmembranenschrumpfungen des Ependyms im Hypothalamus der Ratte, hervorgerufen durch langdauernde Thiobarbiturat-Narkose. *Z. Zellforsch.* **112,** 542–550.

Desaga, U. (1972): Form und Verteilung subependymaler Basalmembranelabyrinthe am Ventrikelsystem der Ratte. *Z. Zellforsch.* **132,** 553–562.

Dierickx, K. (1962): The dendrites of the preoptic neurosecretory nucleus of *Rana temporaria* and the osmoreceptors. *Arch. Int. Pharmacodyn. Ther.* **140,** 708–725.

Fasolo, A., Franzoni, M. F., Mazzi, V. (1972): The neurophysis of the crested newt. II. Fine structure of the pars nervosa with special reference to ependymal cells. *Z. Zellforsch.* **134,** 367–382.

Felix, D., Akert, K. (1977): Peptide action in the cat subfornical organ, In: G. Sterba and W. Bargmann (eds). *Circumventrikuläre Organe.* Nova Acta Leopoldina, Suppl. 9, Halle, 1977. pp. 172–182.

Firth, J. A. (1976): Distribution and properties of an adenosine triphosphate in the tanycyte ependyma of the IIIrd ventricle of the rat. *Histochemistry* **47,** 145–157.

Fleischauer, K. (1957): Untersuchungen am Ependym des zwischen- und mittel-Hirns der Landschildkröte. *Z. Zellforsch.* **46,** 729–767.

Fleischauer, L. (1972): Ependyma and subependymal layer, In: G. H. Bourne (ed.). *The Structure and Function of Nervous Tissue.* Vol. VI. Academic Press, New York, pp. 1–46.

Goslar, H. G., Bock, R. (1971): Histochemische Eigenschaften der unspezifischen Esterasen im Tanycytenependym des III. Ventrikels, im Subfornicalorgan und im Subcommissuralorgan der Wistarratte. *Histochemie* **28,** 170–182.

Güldner, F. H., Wolff, J. R. (1973): Neurono-glial synaptoid contacts in the median eminence of the rat: ultrastructure, staining properties and distribution on tanycytes. *Brain Res.* **61,** 217–234.

Hannah, R. S., Geber, W. (1977): Specializations of the ependyma in the third ventricle of the developing hamster. *Amer. J. Anat.* **149,** 597–603.

Hetzel, W. (1977): Das Ependym der Seitenventrikel von *Acanthodactylus pardalis* (Reptilia, Lacertidae). *Acta Anat. (Basel)* **97,** 68–80.

Hinds, J. W., Ruffett, T. L. (1971): Cell proliferation in the neural tube. An electron microscopic and Golgi analysis in the mouse cerebral vesicle. *Z. Zellforsch.* **115,** 226–264.

Hirano, A., Zimmerman, H. M. (1967): Some new cytological observations of the normal rat ependymal cell. *Anat. Rec.* **158,** 293–302.

Hoffmann, G. E. (1976): Immunocytochemical localization of luteinizing hormone-releasing hormone

(LH-RH) in marine and primate brain. *Anat. Rec.* **184,** 429–430.

HORSTMANN, E. (1954): Die Faserglia des Selachiergehirns. *Z. Zellforsch.* **39,** 588–617.

ITO, H. (1965): The neurosecretory apparatus in the ventricular wall of the reptilian brain. *J. Hirnforsch.* **7,** 493–495.

JACOBS, J. J., MONROE, K. D. (1977): A scanning electron microscopic survey of the brain ventricular system of the female armadillo. *Cell Tissue Res.* **183,** 531–539.

JENNES, V. L., SIKORA, K., SIMONSBERGER, P., ADAM, H. (1977): Rasterelektronmikroskopische Untersuchungen über den Bildungsvorgang des Reissnerschen Fadens bei *Rattus rattus* (L.). *Mikroskopie* **33,** 190–202.

JOSEPH, S. A., SORRENTINO, S., SUNDBERG, D. K. (1975): Releasing hormones, LRF and TRF, in the cerebrospinal fluid of the third ventricle. In: K. M. Knigge, D. E. Scott, H. Kobayashi, S. Ishii (eds). *Brain-Endocrine Interaction II.* Karger, Basel, pp. 306–312.

KARNOVSKY, M. J. (1965): A formaldehyde-glutaraldehyde fixative of high osmolarity for use in electron microscopy. *J. Cell Biol.* **27,** 137–138.

KIRSCHE, W. (1970): Weitere Untersuchungen über das Vorkommen postembryonaler Matrixzonen im Telencephalon einiger Säugetiere. *Z. Mikrosk. Anat. Forsch.* **82,** 122–145.

KISS, A., MITRO, A. (1976): The ependyma of ventriculus mesencephali in golden hamsters. *Anat. Anz.* **140,** 458–466.

KIZER, J. S., PALKOVITS, M., BROWNSTEIN, M. J. (1976): Releasing factors in the circumventricular organs of the rat brain. *Endocrinology* **98,** 311–317.

KLINKERFUSS, G. H. (1964): An electron microscopic study of the ependyma and subependyma glia of the lateral ventricle of the cat. *Amer. J. Anat.* **115,** 71–100.

KNIGGE, K. M., SCOTT, D. E. (1970): Structure and function of the median eminence. *Amer. J. Anat.* **129,** 223–244.

KNIGGE, K. M., JOSEPH, S. A. (1974): Thyrotropin releasing factor (TRF) in cerebrospinal fluid of the third ventricle of rat. *Acta Endocrinol.* (*Kbh.*) **76,** 209–213.

KNIGGE, K. M., SCOTT, D. E., WEINDL, A. (eds). (1972): *Brain-Endocrine Interaction I. Median Eminence: Structure and Function.* Karger, Basel, pp. 1–367.

KNIGGE, K. M., JOSEPH, S. A., SCHOCK, D., SILVERMAN, A. J., CHING, M., SCOTT, D. E., ZEMAN, D., KROBISH-DUDLEY, G. (1973): Role of the ventricular system in neuroendocrine processes: synthesis and distribution of thyrotropin releasing factor (TRF) in hypothalamus and third ventricle. *Can. J. Neurol. Sci.* **1,** 74–84.

KNIGGE, K. M., SCOTT, D. E., KOBAYASHI, H., ISHII, S. (eds). (1975): *Brain-Endocrine Interaction II. The Ventriculus System in Neuroendocrine Mechanisms.* Karger, Basel, pp. 1—395.

KNOWLES, F. (1969): Ependymal secretion, especially in the hypothalamic region. *J. Neuro-Visc. Relat.* **9,** (Suppl.), 97–110.

KNOWLES, F. (1972): Ependyma of the third ventricle in relation to pituitary function. In: *Progress in Brain Res.* Vol. 38. J. Ä. Kappers, J. P. Schadé (eds). *Topics in Neuroendocrinology,* Amsterdam, Elsevier, pp. 255–270.

KNOWLES, F. (1974): Cerebrospinal fluid and endocrine regulation. In: A. Mitro (ed.). *Ependyma and Neurohormonal Regulation.* Veda, Publishing House of the Slovak Academy of Sciences, Bratislava, pp. 11–28.

KOBAYASHI, H., MATSUI, T. (1967): Synapses in the rat and pigeon median eminence. *Endocrinol. Jap.* **14,** 279–283.

KÖLLIKER, A. (1896): *Handbuch der Gewebelehre des Menschen.* 2nd ed., Leipzig, Engelmann.

KOZLOWSKI, G. P., ZIMMERMAN, E. A. (1974): Localization of gonadotropin-releasing hormone (Gn-RH) in sheep and mouse brain. *Anat. Rec.* **178,** 396.

KOZLOWSKI, G. P., FRENK, S. (1977): Ultrastructural localization of gonadotropin-releasing hormone (Gn-RH) in the rat. *Anat. Rec.* **187,** 629.

KOZLOWSKI, G. P., SCOTT, D. E., DUDLEY, G. K. (1973): Scanning electron microscopy of the third ventricle of sheep. *Z. Zellforsch.* **136,** 169–176.

KROON, D. B., GOOSSENS, E. (1974): Protein granules in glial and ependymal cells in the hypophysiotrophic area of the rat, stainable with aldehyde-fuchsin and chrome alum-hematoxylin. *Acta Anat.* (*Basel*) **88,** 267–280.

LEONHARDT, H. (1966): Über ependymale Tanycyten des III. Ventrikels bein Kaninchen in elektronmikroskopischer Betrachtung. *Z. Zellforsch.* **74,** 1–11.

LEONHARDT, H. (1968): Bukettförmige Strukturen im Ependyma der Regio hypothalamica des III. Ventrikels beim Kaninchen. *Z. Zellforsch.* **88,** 297–317.

LEONHARDT, H. (1969): Ependym. In: G. Sterba (ed.), *Zirkumventrikuläre Organe und Liquor.* Fischer Verlag, Jena, pp. 177–190.

LEONHARDT, H. (1972): Über die topographische Verteilung des subependymalen Basalmembranlabyrinthe im Ventrikelsystem des Kaninchengehirns. *Z. Zellforsch.* **127,** 392–406.

LEONHARDT, H. (1974): Ependym Strukturen im Dienst des Stofftransportes zwischen Ventrikelliquor und Hirnsubstanz. In: A. Mitro (ed.), *Ependyma and Neurohormonal Regulation.* Veda, Publishing House of the Slovak Academy of Sciences, Bratislava, pp. 29–75.

LEONHARDT, H., LINDNER, E. (1967): Marklose Nervenfasern im III. und IV. Ventrikel des Kaninchen- und Katzenhirns. *Z. Zellforsch.* **78,** 1–18.

LEONHARDT, H., PRIEN, H. (1968): Eine weitere Art intraventrikulär kolbenförmiger Axonendigungen aus dem IV. Ventrikel des Kaninchengehirns. *Z. Zellforsch.* **92,** 394–399.

Leonhardt, H., Backhus-Roth, A. (1969): Synapsenartige Konktakte zwischen intraventrikulären Axonendigungen und freien Oberflächen von Ependymzellen des Kaninchengehirns. *Z. Zellforsch.* **97**, 369–376.

Leonhardt, H., Desaga, U. (1975): Recent observations on ependymal and subependymal basement membranes. *Acta Neurochir.* (*Wien*) **31**, 153–159.

Léránth, Cs., Schiebler, T. H. (1974): Über die Aufnahme von Peroxidase aus dem III. Ventrikel der Ratte. *Brain Res.* **67**, 1–11.

Levinger, I. M. (1971): The cerebral ventricles of the rat. *J. Anat.* **108**, 447–451.

Levinger, I. M., Kedem, J. (1974): A method for the evaluation of the surface area of cerebral ventricles in animals. *J. Anat.* **117**, 481–485.

Lewis, P. R., Shute, C. C. D. (1967): The cholinergic limbic system: projections to hippocampal formation, medial cortex, nuclei of ascending cholinergic reticular system, and the subfornical organ and supra-optic crest. *Brain* **90**, 521–540.

Lindberg, L. A., Vesenius, L., Talanti, S. (1977): The surface fine structure of the ependymal lining of the lateral ventricle in rats with hereditary hydrocephalus. *Cell Tissue Res.* **179**, 121–129.

Löfgren, F. (1959a): The infundibular recess, a component in the hypothalamus-adenohypophyseal system. *Acta Morphol. Neerl. Scand.* **3**, 55–78.

Löfgren, F. (1959b): New aspects of the hypothalamic control of the adenohypophysis. *Acta Morphol. Neerl. Scand.* **2**, 220–229.

Lorez, H. P., Richards, J. G. (1973): Distribution of indolealkylamine nerve terminals in the ventricle of the rat brain. *Z. Zellforsch.* **144**, 511–522.

Lorez, H. P., Richards, J. G. (1975): 5-HT nerve terminals in the fourth ventricle of the rat brain: their identification and distribution studied by fluorescence histochemistry and electron microscopy. *Cell Tissue Res.* **165**, 37–48.

Lorez, H. P., Pieri, L., Richards, J. G. (1975): Disappearance of supra-ependymal 5-HT axons in the rat forebrain after electrolytic and 5.6-DHT-induced lesions of the medial forebrain bundle. *Brain Res.* **100**, 1–12.

Luppa, H., Feustel, G. (1971): Location and characterization of hydrolytic enzymes of the third ventricle lining in the region of the recessus infundibularis of the rat. A study on the function of the ependyma. *Brain Res.* **29**, 253–270.

McNeill, T. H. (1977): The comparative immunocytochemical localization of neurophysin and gonadotropin releasing hormone in the mammalian and avian organum vasculosum laminae terminalis. *Anat. Rec.* **187**, 652.

Makara, G. B., Palkovits, M., Szentágotai, J. (1980): The endocrine hypothalamus and the hormonal response to stress. In: H. Selye (ed.). *Selye's Guide to Stress Research.* Vol. 1. Van Nostrand Reinhold, New York, etc. pp. 280–337.

Malinský, J. (1968): Fine structure of ependyma in lateral ventricles of human brain. *Acta Univ. Palackianae Olomucensis* **48**, 65–72.

Manthorpe, M. C., Wilkin, G. P., Wilson, J. E. (1977): Purification of viable ciliated cuboidal ependymal cells from the rat brain. *Brain Res.* **134**, 407–415.

Marker, G. (1970): Fasergliastruktur der dorsalen Wand des Aquaeductus Cerebri bei einigen Primaten. *Z. Zellforsch.* **107**, 564–585.

Martínez Martínez, P., Weerd de, H. (1977): The fine structure of the ependymal surface of the recessus infundibularis in rat. *Anat. Embryol.* (*Berl.*) **151**, 241–265.

Matano, S., Ishii, S. (1973): The fine structural localization of Na-K ATP-ase in the choroid plexus in rat. *J. Electron Microsc.* (*Tokyo*) **22**, 357–359.

Matsui, T. (1966): Fine structure of the median eminence of the rat. *J. Fac. Sci. Univ. of Tokyo* **11**, 71–96.

Mesters, P., Breipohl, W. (1976): Morphology and distribution of subependymal cells in the third ventricle of the albino rat. *Cell Tissue Res.* **168**, 303–314.

Mészáros, T., Csuri, I. J., Házas, J., Palkovits, M. (1969): Esterase activity in the hypothalamus. *Acta Morphol. Acad. Sci. Hung.* **17**, 201–215.

Mészáros, T., Léránth, Cs., Palkovits, M., Házas, J. (1969): Secretory and esterase activity of the circumventricular organs with special reference to the infundibular ependyma. In: G. Sterba (ed.), *Zirkumventrikuläre Organe und Liquor.* Fischer, Jena, pp. 131–134.

Milhorat, T. H., Hammock, M. K., Fenstermacher, J. D., Rall, D. P., Levin, V. A. (1971): Cerebrospinal fluid production by the choroid plexus and brain. *Science* **173**, 330–332.

Millhouse, O. E. (1971): A Golgi study of third ventricle tanycytes in the adult rodent brain. *Z. Zellforsch.* **121**, 1–13.

Millhouse, O. E. (1972): Light and electron microscopic studies of the ventricular wall. *Z. Zellforsch.* **127**, 149–174.

Mitro, A. (1969): Über ein spezielles Ependym im III. Ventrikel der Ratte. *Experientia* **25**, 287.

Mitro, A. (ed.) (1974): *Ependyma and Neurohormonal Regulation.* Veda, Publishing House of the Slovak Academy of Sciences, Bratislava, pp. 1–323.

Mitro, A. (1975): Light microscopy and histochemical study of the ependyma of the third cerebral ventricle in the albino rat. *Folia morphol.* (*Praha*) **4**, 347–356.

Mitro, A. (1976): *Ependyma of the Rat Brain Ventricles.* Veda, Publishing House of the Slovak Academy of Sciences, Bratislava, pp. 1–145 (In Slovak).

Mitro, A., Schiebler, T. H. (1972): Über Entwicklung regionaler Unterschiede im Ependym des III. Ventrikels der Ratte. *Anat. Anz.* **132**, 1–19.

Mitro, A., Kiss, A. (1977): Histological observations on the ependyma of the ventriculus mesencephali in the guinea pig. *Acta Anat.* (*Basel*) **97**, 248–256.

Moss, M. L., Mechan, M. A., Salentijn, L. (1972): Transformative and translative growth processes in neurocranial development of the rat. *Acta Anat.* (*Basel*) **81**, 161–182.

NAKAI, Y. (1971): Fine structure and its functional properties of the ependymal cell in the frog median eminence. *Z. Zellforsch.* **122**, 15–25.

NAKAI, Y., OCHIAI, H., SHIODA, S. (1977a): Cytological evidence for different types of cerebral fluid-contacting subependymal cells in the preoptic and infundibular recesses of the frog. *Cell Tissue Res.* **176**, 317–334.

NAKAI, Y., OCHIAI, H., UCHIDA, M. (1977b): Fine structure of ependymal cells in the median eminence of the frog and mouse revealed by freeze-etching. *Cell Tissue Res.* **181**, 311–318.

NAKAYAMA, Y., KOHNO, K. (1974): Number and polarity of the ependymal cilia in the central canal of some vertebrates. *J. Neurocytol.* **3**, 449–458.

NOACK, W., WOLFF, J. R. (1970): Über neuritenähnliche intraventrikuläre Forsätze und ihre Kontakte mit dem Ependym der Seitenventrikel der Katze. *Z. Zellforsch.* **11**, 572–585.

NOACK, W., DUMITRESCU, L., SCHWEICHEL, J. U. (1972): Scanning and electron microscopical investigations of the surface structures of the lateral ventricles in the cat. *Brain Res.* **46**, 121–129.

OGATA, J., HOCHWALD, G. M., CRAVIOTO, H., RANSCHOFF, J. (1972): Light and electron microscopic studies of experimental hydrocephalus. Ependymal and subependymal areas. *Acta Neuropathol.* (*Berl.*) **21**, 213–223.

OOTA, A., KOBAYASHI, H., NISHIOKA, R. S., BERNA, H. A. (1974): Relationship between neurosecretory axon and ependymal terminals on capillary walls in the median eminence of several vertebrates. *Neuroendocrinology* **16**, 127–136.

PALKOVITS, M. (1965): *Morphology and Function of the Subcommissural Organ.* Akadémiai Kiadó, Budapest, pp. 1–105.

PALKOVITS, M. (1978): Neurochemical anatomy of the neuroendocrine hypothalamus. *Bull. Schweiz. Akad. Med. Wiss.* **34**, 113–130.

PALKOVITS, M., MAGYAR, P., SZENTÁGOTHAI, J. (1971): Quantitative histological analysis of the cerebellar cortex in the cat. I. Number and arrangement in space of the Purkinje cells. *Brain Res.* **32**, 1–13.

PALKOVITS, M., BROWNSTEIN, M., ARIMURA, A., SATO, H., SCHALLY, A. V., KIZER, J. S. (1976): Somatostatin content of the hypothalamic ventromedial and arcuate nuclei and the circumventricular organs in the rat. *Brain Res.* **109**, 430–434.

PALKOVITS, M., MEZEY, É., AMBACH, G., KIVOVICS, P. (1978): Neural and vascular connections between the organum vasculosum laminae terminalis and preoptic nuclei. In: D. E. Scott, G. P. Kozlowski, A. Weindl (eds). *Brain-Endocrine Interaction III. Neural Hormones and Reproduction.* Karger, Basel, pp. 302–313.

PAUL, W. K., MARTIN, H., SCOTT, D. E. (1977): Scanning electron microscopy of the third ventricular floor of the rat. *J. Comp. Neurol.* **176**, 301–310.

PESONEN, N. (1940): Über die intraependymalen Nervenelements. *Anat. Anz.* **90**, 193–223.

PETERS, A. (1974): The surface fine structure of the choroid plexus and ependymal lining of the rat lateral ventricle. *J. Neurocytol.* **3**, 99–108.

PETERS, A., PALAY, S. L., WEBSTER, H. DE F. (1976): *The Fine Structure of the Nervous System. The Neurons and Supporting Cells.* Saunders, Philadelphia—London—Toronto, pp. 1–406.

PHILLIPS, M. I., BALHORN, L., LEAVITT, M., HOFFMAN, W. (1974): Scanning electron microscopic study of the rat subfornical organ. *Brain Res.* **80**, 95–110.

PRIVAT, A. (1977): The ependyma and subependymal layer of the young rat. A new contribution with freeze-fracture. *Neuroscience* **2**, 447–457.

PRIVAT, A., LEBLOND, C. P. (1972): The subependymal layer and neighbouring region in the brain of the young rat. *J. Comp. Neurol.* **146**, 277–302.

PURKINJE, J. (1836): Über Flimmerbewegungen im Gehirn. *Arch. Anat. Physiol. Wiss. Med.* (*Berl.*) **3**, 289–290.

REYNOLDS, E. S. (1963): The use of lead citrate at high pH as an electronopaque stain in electron microscopy. *J. Cell Biol.* **17**, 208–212.

RIBAS, J. L. (1977a): Morphological evidence for a possible functional role of supra-ependymal nerves on ependyma. *Brain Res.* **125**, 362–368.

RIBAS, J. L. (1977b): The rat epithalamus. I. Correlative scanning-transmission electron microscopy of supraependymal nerves. *Cell Tissue Res.* **182**, 1–16.

RINNE, U. K. (1966): Ultrastructure of the median eminence of the rat. *Z. Zellforsch.* **74**, 98–122.

RODRIGUEZ, E. M. (1969): Ependymal specializations. I. Fine structure of the neural region of the toad median eminence, with particular reference to the connections between the ependymal cells and the subependymal capillary loops. *Z. Zellforsch.* **102**, 153–171.

RODRIGUEZ, E. M. (1970a): Ependymal specializations. II. Ultrastructural aspect of the apical secretion of the subcommissural organ. *Z. Zellforsch.* **111**, 15–31.

RODRIGUEZ, E. M. (1970b): Ependymal specializations. III. Ultrastructural aspects of the basal secretion of the toad subcommissural organ. *Z. Zellforsch.* **11**, 32–50.

RÖHLICH, P., VIGH, B., TEICHMANN, I., AROS, B. (1965): Electron microscopy of the median eminence of the rat. *Acta Biol. Acad. Sci. Hung.* **15**, 431–457.

SAAVEDRA, J. M., BROWNSTEIN, M., KIZER, J. S., PALKOVITS, M. (1976): Biogenic amines and related enzymes in the circumventricular organs of the rat. *Brain Res.* **107**, 412–417.

SALAND, L. C., EVAN, A. P., DEMSKI, L. S. (1974): Ultrastructre of ependymal cells of the shark median eminence. *Anat. Rec.* **178**, 657–666.

SARNAT, H. B., CAMPA, J. F., LLOYD, J. M. (1975): Inverse prominence of ependyma and capillaries in the spinal cord of vertebrates. A comparative histochemical study. *Amer. J. Anat.* **143**, 439–450.

SCHACHENMAYR, W. (1967): Über die Entwicklung von Ependym und Plexus choroideus der Ratte. *Z. Zellforsch.* **77**, 25–63.

SCHIEBLER, T. H., MITRO, A. (1969): Über die Entwicklung des Ependymas, In: G. Sterba (ed.), *Zirkumventrikuläre Organe und Liquor*. Fischer, Jena, pp. 219–222.

SCHIMRIGH, K. (1966): Über die Wandstruktur der Seitenventrikel und des dritten Ventrikels beim Menschen. *Z. Zellforsch.* **70**, 1–20.

SCHWANITZ, W. (1969): Die topographische Verteilung supraependymaler Strukturen in den Ventrikeln und im Zentralkanal des Kaninchengehirns. *Z. Zellforsch.* **100**, 536–551.

SCOTT, D. E., PAULL, K. W., KROBISCH-DUDLEY, G. (1972): A comparative scanning microscopic analysis of the human cerebral ventricular system. I. The third ventricle. *Z. Zellforsch.* **132**, 203–215.

SCOTT, D. E., KOZLOWSKI, G. P., PAULL, W. K., RAMALINGAM, S., DUDLEY, G. K. (1973): Scanning electron microscopy of the human cerebral ventricular system. II. The fourth ventricle. *Z. Zellforsch.* **139**, 61–68.

SCOTT, D. E., KOZLOWSKI, G. P., SHERIDAN, M. N. (1974a): Scanning electron microscopy in the ultrastructural analysis of the mammalian cerebral ventricular system. *Int. Rev. Cytol.* **37**, 349–388.

SCOTT, D. E., KROBISCH-DUDLEY, G., KNIGGE, K. M. (1974b): The ventricular system in neuroendocrine mechanisms. II. *In vivo* monoamine transport by ependyma of the median eminence. *Cell Tissue Res.* **154**, 1–16.

SCOTT, D. E., KROBISCH-DUDLEY, G., KNIGGE, K. M., KOZLOWSKI, G. P. (1974c): *In vitro* analysis of the cellular localization of the luteinizing hormone-releasing factor (LRF) in the basal hypothalamus of the rat. *Cell Tissue Res.* **149**, 371–378.

SHIODA, S., HONMA, Y., YOSHIDA, S. (1977): Scanning electron microscopy of the third ventricular wall in the lamprey. *Arch. Histol. Jap.* **40**, 41–49.

SHUTTLEWORTH, E. C., ALLEN, N. (1966): Acid hydrolases in pia-arachnoid and ependyma of the rat brain. *Neurology (Minneap.)* **16**, 979–985.

SILVERMAN, A. J., KNIGGE, K. M., RIBAS, J. L., SHERIDAN, M. R. (1973): Transport capacity of median eminence. III. Amino acid and thyroxine transport of organ-cultured median eminence. *Neuroendocrinology* **11**, 107–118.

SILVERMAN, A. J. ANTUNES, J. L., FERIN, M., ZIMMERMAN, E. A. (1977): The distribution of luteinizing hormone-releasing hormone (LH-RH) in the hypothalamus of the rhesus monkey. Light microscopic studies using an immunoperoxidase technique. *Endocrinology* **101**, 134–141.

SOEMMERING, S. T. (1841): Hirn- und Nervenlehre. (Umgearbeitet von G. Valentin.) L. Voss, Leipzig.

ŠPAČEK, J., PAŘIZEK, J. (1969): The fine structure of the area postrema of the rat. *Acta Morphol. Acad. Sci. Hung.* **17**, 17–34.

SREBRO, Z. (1967): Observations on the cytology of frog brain ependyma. *Folia Biologica (Warsava)* **3**, 265–274.

STERBA, G. (ed.) (1969): *Zirkumventrikuläre Organe und Liquor*. Fischer, Jena, pp. 1–309.

STERBA, G. (1977): Das Subkommissuralorgan. In: G. Sterba and W. Bargmann (eds), *Zirkumventrikuläre Organe*. Nova Acta Leopoldina, Suppl. 9, Halle, pp. 103–114.

STERBA, G. and BARGMANN, W. (eds) (1977): *Zirkumventrikuläre Organe*. Nova Acta Leopoldina, Suppl. 9, Halle pp. 1–252.

STUDNIČKA, F. K. (1900): Untersuchungen über den Bau des Ependyma der nervösen Zentralorgane. *Anat. Hefte* **115**, 303–430.

TAKEICHI, M. (1966): The fine structure of ependymal cells. Part I. The fine structure of ependymal cells in the kitten. *Arch. Histol. Jap.* **26**, 483–505.

TENNYSON, V. M., PAPPAS, G. D. (1962): An electron microscopic study of ependymal cells of the fetal, early postnatal and adult rabbit. *Z. Zellforsch.* **56**, 595–618.

TERNEAUX, J. P., HENRY, F., HAMON, M., BOURGOIN, S., GLOWINSKI, J. (1977): 5-HT release from ependymal surface of the caudate nucleus in "encéphale isolé" cats. *Brain Res.* **132**, 575–579.

UEMURA, H. and KOBAYASHI, H. (1977): Effect of gonadal function by lesioning tanycytes in median eminence of the rat and Japanese gail. *Cell Tissue Res.* **178**, 143–153.

UEMURA, H., ASAI, T., NOZAKI, M., KOBAYASHI, H. (1975): Ependymal absorption of luteinizing hormone releasing hormone injected into the third ventricle of the rat. *Cell Tissue Res.* **160**, 443–452.

VERHOEF, J., PALKOVITS, M., WITTER, A. (1977): Distribution of a behaviorally highly potent ACTH 4-9 analog in rat brain after intraventricular administration. *Brain Res.* **126**, 89–104.

VIGH, B. (1971): *Das Paraventrikularorgan und das zirkumventrikuläre System des Gehirns*. Akadémiai Kiadó, Budapest, pp. 1–149.

VIGH, B., VIGH-TEICHMANN, I. (1973): Comparative ultrastructure of the cerebrospinal fluid-containing neurons. *Int. Rev. Cytol.* **35**, 189–251.

VIGH-TEICHMANN, I., VIGH, B. (1974): The infundibular cerebrospinal-fluid contacting neurons. *Adv. Anat. Embryol. Cell Biol.* **50**, pp. 1–91.

WAGNER, H. J., PILGRIM, CH. (1974): Extracellular and transcellular transport of horseradish peroxidase (HRP) through the hypothalamic tanycyte ependyma. *Cell Tissue Res.* **152**, 477–491.

WEINDL, A. (1973): Neuroscience aspects of circumventricular organs. In: W. F. Ganong and L. Martini (eds), *Frontiers in Neuroendocrinology*. Oxford University Press, London, pp. 3–32.

WEINDL, A., SOFRONIEW, M. V. (1978): Neurohormones and circumventricular organs. An immunohistochemical investigation. In: D. E. Scott, G. P. Kozlowski, A. Weindl (eds), *Brain-Endocrine Interaction III. Neural Hormones and Reproduction.* Karger, Basel, pp. 117–137.

Weindl, A., Schinko, I. (1975): Vascular and ventricular neurosecretion in the organum vasculosum of the golden hamster. In: K. M. Knigge, D. E. Scott, H. Kobayashi, S. Ishhi (eds), *Brain-Endocrine Interaction II.* Karger, Basel, pp. 190–203.

Weindl, A., Schwink, A., Wetzstein, R. (1967): Der Feinbau des Gefässorgans der Lamina terminalis beim Kaninchen. I. Die Gefässe. *Z. Zellforsch.* **79,** 1–48.

Weindl, A., Schwink, A., Wetzstein, R. (1968): Der Feinbau des Gefässorgans der Lamina terminalis beim Kaninchen. II. Das neuronale und gliale Gewebe. *Z. Zellforsch.* **85,** 552–600.

Weindl, A., Schinko, I., Sofroniew, M. V. (1978): Ultrastructural, fluorescence and immunohistochemical investigations of the organum vasculosum of the lamina terminalis. In: W. Bargman, Oksche, A. Polenov, B. Scharrer (eds). *Neuroendocrine Activity. Evolution, Structure and Function.* Springer, Berlin—Heidelberg, pp. 319.

Weindl, A., Sofroniew, M. V. (1978): Neurohormones and circumventricular organs. An immunohistochemical investigation. In: D. E. Scott, G. P. Kozlowski, A. Weindl (eds), *Brain-Endocrine Interaction III. Neural Hormones and Reproduction.* Karger, Basel, pp. 117–137.

Wenger, T. (1976): Ultrastructural changes in the nerve terminals of the vascular organ of the lamina terminalis in the rat during the estrous cycle. *Neuroscience Letters* **3,** 29–32.

Wenger, T. (1977): The effect of ovariectomy on the ultrastructure of the organon vasculosum laminae terminalis in the rat, In: G. Sterba and W. Bargmann (eds), *Circumventrikuläre Organe.* Nova acta Leopoldina, Suppl. 9, Halle, pp. 217–220.

Westergaard, E. (1969): The cerebral ventricles of the rat during growth. *Acta Anat.* (*Basel*) **74,** 405–423.

Westergaard, E. (1970): *The Lateral Cerebral Ventricles and the Ventricular Walls.* Andelsbogtrykkeriet, Odense, pp. 1–216.

Westergaard, E. (1972): The fine structure of nerve fibre and endings in the lateral cerebral ventricles of the rat. *J. Comp. Neurol.* **144,** 345–354.

Wittkowski, W. (1968): Zur funktionellen Morphologie ependymaler und extraependymaler Glia im Rahmen der Neurosekretion. Elektronmikroskopische Untersuchungen an der Neurohypophyse der Ratte. *Z. Zellforsch.* **86,** 111–128.

Wittkowski, W. (1972): Zur Ultrastruktur der Gefässfortsätze von Ependyma und Gliazellen im Infundibulum der Ratte. *Z. Zellforsch.* **130,** 58–59.

Wittkowski, W. (1973): Elektronmikroskopische Untersuchungen zur funktionellen Morphologie des tuberohypophysären Systems der Ratte. *Z. Zellforsch.* **139,** 101–148.

Worthington, W. C. Jr., Catchcart, R. S. (1963): Ependymal cilia: distribution and activity in the adult human brain. *Science* **139,** 221–222.

Yamadori, T. (1977): The directions of ciliary beat on the wall of the fourth ventricle in the mouse. *Arch. Histol. Jap.* **40,** 283–296.

Yamadori, T., Yagihashi, S. (1975): A scanning and transmission electron microscopic observation of the fourth ventricular floor in the mouse. *Arch. Histol. Jap.* **37,** 415–432.

Zimmerman, E. A., Antunes, J. L. (1976): Organization of the hypothalamic-pituitary system: current concepts from immunocytochemical studies. *J. Histochem. Cytochem.* **24,** 807–815.

Zimmerman, E. A., Hsu, K. C., Ferin, M., Kozlowski, G. B. (1974): Localization of gonadotropin-releasing hormone (Gn-RH) in the hypothalamus of the mouse by immunoperoxidase technique. *Endocrinology* **95,** 1–8.

SUBJECT INDEX

amygdala (A) 14
area postrema (ap) 28–29, 93
area praetectalis (apt) 17
area retrochiasmatica (ar) 15, 57–58
arcuate-tanycyte zone 66–67
"arcuate-type" tanycyte (inframamillary recess) 79, 82
A2-catecholaminergic cell group (A2) 29

basal membrane labyrinths 63

capsula externa (CAE) 14, 17–18, 47
capsula interna (CAI) 14–15, 17
central canal (medullary part) (ca) 7–8, 27, 29, 94–98
central part of the lateral ventricle (cL) 14
cerebral aqueduct (AC) 7, 8, 14, 16, 18, 20, 84–90, 93
  caudal ventral wall 85–86, 89
  dorsal cylindric ependyma 84–86
  dorsal wall in the central part 85–88
  ventral cubic ependyma 84–86
  ventral wall in the central part 85–86, 88–89
chiasma opticum (CO) 13, 54, 57
choroid plexus 35, 53, 56, 59
cilia 38, 61–62, 92
cingulum (C) 11, 13
circumventricular organs (CVO) 8, 39
colliculus inferior (CI) 7–8, 16, 20, 22–23, 32, 86
colliculus superior (CS) 18, 32, 86
commissura anterior (CA) 10, 12–13, 33, 54, 58
commissura fornicis ventralis (CFV) 15
commissura posterior (CP) 7, 16, 19, 87
cornu anterious of the lateral ventricle (ca) 10, 12
cornu inferius of the lateral ventricle (ci) 16, 33
corpus callosum (CC) 10, 12, 31–32, 43
corpus geniculatum laterale (cgl) 17
CSF-contacting neurons 39
C2 adrenaline-containing cell group (C2) 27

decussatio pyramidis (DP) 28
double-layered tanycyte zone 66–67, 68

eminentia mediana (median eminence, me) 17, 48, 54–56, 58, 65–67, 78, 100
  postinfundibular part 79
ependyma, development 36–37
ependymal astrocytes 37
ependymal tanycytes 37
epithalamus 86

fasciculus gracilis (FG) 29
fasciculus longitudinalis medialis (FLM) 21, 25–27
fasciculus retroflexus (FR) 16
fastigium (f) 7
fimbria hippocampi (FH) 14–15, 17, 32, 45–47
fimbria septi (FS) 12–13
fornix (F) 13
fourth ventricle (IVth) 7–9, 22–23, 26, 28, 33, 86, 90–98
  choroidea ventriculi IV. 91, 94
  cuboidal ependyma 91, 94, 98
  ependyma on the roof 91, 94–96
  flattened ependyma 91, 94
  lamina epithelialis of the tela 91, 94, 97
  ventral wall 91, 94, 97
freeze-etching technique 36

globus pallidus (gp) 13
gyrus dentatus (GD) 15–18, 20

habenula (H) 16, 32
hippocampus (HI) 15–18, 31–33

inframamillary recess (RI) 7, 16, 48, 54–56, 58, 74, 76, 78–79, 81–84, 100
interventricular foramen (FI) 9, 12, 30, 33, 48–51
intraependymal cisterna 62
intraventricular nerve fibres 38, 61, 96
irregular tanycyte zone 66–67, 68

lamina epithelialis (IVth ventricle) 97
lateral recess of the IVth ventricle (RL) 24, 95
lateral ventricle (LV) 7–8, 39–40, 42–44
  anterior horn 7, 42–44
    anterior area 40, 42
    external wall 42–43
    top ependymal area 40, 42
    transitional area 42, 44
    upper internal wall 40, 42–43
  central part 7, 40–42, 44–45
    caudal portion 40–42, 45
      external wall 40–42, 45
      internal wall 40–42, 45
    middle portion 40–42, 44
      external wall 40–42, 45–46
      internal wall 40–52, 44–45
    inferior horn 7, 16–17, 41–42, 47
locus coeruleus (lc) 23

mamillary body (MB) 16, 54, 84
massa intermedia of the thalamus (mi) 7, 31, 33, 54
medulla oblongata (MO) 31
mesencephalic ependyma 84–85, 87–89
mesencephalic ventricle 7–8, 20, 22, 84–87, 89–92
microvilli 37, 61–62, 92
monolayered tanycyte ependyma 66–68
monolayered tanycyte zone 66–68
multilayered ependyma 66–68

nervus facialis (VII.) 24–25
nuclei lineares (nl) 19
nucleus accessorius (Darkschewitch, aD) 19
nucleus accumbens (a) 11
nucleus anterior dorsalis thalami (tad) 15
nucleus anterior ventralis thalami (tav) 15
nucleus arcuatus (na) 15, 17, 66–68, 78–79, 100
nucleus caudatus (c) 10–11, 13, 32
nucleus caudatus–putamen (cp) 12–13, 16–17, 33
nucleus cochlearis dorsalis (cod) 25
nucleus commissuralis (nco) 29
nucleus dorsomedialis (ndm) 15, 17
nucleus gracilis (gr) 29
nucleus habenulae lateralis (hl) 17
nucleus habenulae medialis (hm) 17, 50
nucleus hypothalamicus anterior (nha) 15, 58
nucleus hypothalamicus posterior (nhp) 17, 78, 100
nucleus intercalatus (nic) 29
nucleus interpositus cerebelli 25
nucleus interstitialis striae terminalis (nist) 13
nucleus lateralis cerebelli (l) 25
nucleus lateralis posterior thalami (tlp) 17
nucleus lateralis thalami (tl) 15, 17
nucleus mamillaris lateralis (nml) 17
nucleus mamillaris medialis (nmm) 17
    pars lateralis (nmx) 17
    pars medialis 17
nucleus medialis cerebelli (m) 25
nucleus medialis thalami (tm) 15
nucleus motorius nervi hypoglossi (nXII) 29
nucleus originis dorsalis vagi (nX) 29
nucleus parafascicularis (npf) 78
nucleus parataenialis thalami (npv) 13
nucleus paraventricularis (ptv) 13, 15, 50, 58–60, 66–67
nucleus periventricularis hypothalami (npe) 13
nucleus periventricularis thalami (ptv) 17
nucleus praemamillaris dorsalis (npmd) 17, 100
nucleus praeopticus medialis (pom) 13, 58
nucleus praeopticus periventricularis (pop) 13, 58
nucleus praeopticus suprachiasmatis (pos) 13
nucleus praepositus hypoglossi (ph) 27
nucleus raphe dorsalis (rd) 21, 23
nucleus reticularis gigantocellularis (rgi) 27
nucleus reticularis paramedianus (nrp) 27
nucleus reticularis pontis caudalis (rpoc) 25
nucleus reticularis thalami (tr) 15
nucleus reuniens (re) 15, 17, 58
nucleus salivatorius (sa) 25
nucleus septalis dorsalis (sd) 11, 13, 45
nucleus septalis fimbrialis (sf) 13
nucleus septalis intermedius (si) 13
nucleus septalis lateralis (sl) 11, 45
nucleus septalis medialis (sm) 11
nucleus septalis triangularis (st) 13
nucleus suprachiasmatis (nsc) 13, 15
nucleus suprageniculatus (s) 25
nucleus tegmenti dorsalis (ntd) 23, 25
nucleus tegmenti dorsolateralis (ntdl) 23
nucleus tractus diagonalis (td) 13
nucleus tractus solitarii (nts) 27–29
nucleus ventromedialis (nvm) 15, 17, 66–67
nucleus vestibularis lateralis (Deiters) (vl) 25
nucleus vestibularis medialis (vm) 25, 27
nucleus vestibularis superior (vs) 25
nucleus vestibularis spinalis (vsp) 25, 27

obex (O) 28–29, 93
organum vasculosum laminae terminalis (ovlt) 13, 54, 56–57

"paraventricular organ" 71
pars lateralis of the nucleus medialis thalami (tml) 15
pedunculus cerebellaris superior (PCS) 23, 25
pineal gland (PG) 20, 86
pituitary stalk (ps) 7, 54, 78, 100
pons (P) 31
posterior commissure 9
posterior portion of the lateral ventricle (pp) 16, 18

quantitative histology 5

"recess organ" 48, 54–56, 58, 64, 66–67
recessus inframamillaris (RI) 7, 16, 54, 78, 81–83, 100
recessus infundibuli (Ri) 7, 54
recessus opticus (RO) 7, 12, 48, 54–58
reccessus pinealis (RP) 7, 16, 18, 32, 48–51, 53, 76, 86–87
recessus triangularis (RT) 9, 12, 33, 48–51
Reissner's fibre 87

scanning electron microscopy 35–36
septum (SE) 10, 31, 33, 43
stria medullaris thalami (SM) 13, 15, 17
subcommissural organ (SCO) 7, 17, 85–87
subependyma 38, 44
subfornical organ (sfo) 12–13
subiculum (S) 16–17, 19–20, 53
substantia grisea centralis (SGC) 18–21, 23, 31, 88
    pars dorsalis (SGCd) 21
    pars lateralis (SGCl) 19, 21
    pars ventralis (SGCv) 19
supraependymal cells 35–37, 61, 63
supraependymal nerve fibres 61–62
supraependymal structure 37
supraoptic recess 7, 12, 48, 54–58

tanycyte ependyma 48, 54–56, 58, 66–67, 69–70, 72–73, 76, 79, 81, 83
tanycytes 65–67, 69–75, 100
tegmentum pontis (TP) 24

tela choroidea 53, 56, 93, 97
thalamus (TH) 14
third ventricle (IIIrd) 7–9, 48–84, 86, 100
dorsal part (dT) 12, 14, 16, 48–50
epithalamic ependyma 48–52
floor 48–52
side wall 48–52
pineal recess (RP) 7, 16, 18, 32, 48–51, 53, 76, 86–87
triangular recess (RT) + interventricular foramen (FI) 9, 12, 33, 48–51
ventral part (vT) 12, 14, 48, 54–56
anterior hypothalamic portion 48, 54–59, 61, 63
inframamillary recess (RI) 7, 16, 48, 54–56, 58, 74, 76, 78–79, 81–84, 100
roof of the recess 48, 54–56, 58
tanycyte ependyma 48, 54–56, 58, 66–67, 69–70, 72–73, 76, 79, 81, 83
supraoptic recess 48, 54–58
floor 49, 54–56, 58
side wall 48, 54–56, 58
tuberal portion 48, 54–56, 58, 66–67, 77, 84
cuboidal ependyma 48, 54–56, 58, 66–67
"light cell zone" 48, 54–56, 58, 66–67, 73–76
"recess organ" 48, 54–56, 58, 64, 66–67
roof of the median eminence 48, 54–56, 58, 65–67
tanycyte ependyma 58, 64–66, 68, 76–77, 69–70, 72–73
"connecting part" (cT) 12, 16, 33, 48, 54–56, 58, 74, 78–79, 84
tractus diagonalis (TD) 12
tractus opticus (TO) 15
tractus solitarius (TS) 27, 29
tractus striohypothalamicus (TSTH) 11

velum medullare superius (VMS) 93, 95–96
ventriculus mesencephali (VM) 7–8, 20, 22, 84–87, 89–92
vermis cerebelli (V) 23–26, 29, 96